KU-493-869

Techniques for
Engineering Genes

LEEDS BECKETT UNI
Leeds Metropolitan University

17 0093726 8

BOOKS IN THE BIOTOL SERIES

The Molecular Fabric of Cells
Infrastructure and Activities of Cells

Techniques used in Bioproduct Analysis
Analysis of Amino Acids, Proteins and Nucleic Acids
Analysis of Carbohydrates and Lipids

Principles of Cell Energetics
Energy Sources for Cells
Biosynthesis and the Integration of Cell Metabolism

Genome Management in Prokaryotes
Genome Management in Eukaryotes

Crop Physiology
Crop Productivity

Functional Physiology
Cellular Interactions and Immunobiology
Defence Mechanisms

Bioprocess Technology: Modelling and Transport Phenomena
Operational Modes of Bioreactors

In vitro Cultivation of Micro-organisms
In vitro Cultivation of Plant Cells
In vitro Cultivation of Animal Cells

Bioreactor Design and Product Yield
Product Recovery in Bioprocess Technology

Techniques for Engineering Genes
Strategies for Engineering Organisms

Principles of Enzymology for Technological Applications
Technological Applications of Biocatalysts
Technological Applications of Immunochemicals

Biotechnological Innovations in Health Care

Biotechnological Innovations in Crop Improvement
Biotechnological Innovations in Animal Productivity

Biotechnological Innovations in Energy and Environmental Management

Biotechnological Innovations in Chemical Synthesis

Biotechnological Innovations in Food Processing

Biotechnology Source Book: Safety, Good Practice and Regulatory Affairs

BIOTECHNOLOGY BY OPEN LEARNING

Techniques for Engineering Genes

PUBLISHED ON BEHALF OF :

Open universiteit   and   **University of Greenwich (formerly Thames Polytechnic)**

Valkenburgerweg 167
6401 DL Heerlen
Nederland

Avery Hill Road
Eltham, London SE9 2HB
United Kingdom

Butterworth-Heinemann Ltd
Linacre House, Jordan Hill, Oxford OX2 8DP

A member of the Reed Elsevier group

OXFORD LONDON BOSTON
MUNICH NEW DELHI SINGAPORE SYDNEY
TOKYO TORONTO WELLINGTON

First published 1993

© Butterworth-Heinemann Ltd 1993

All rights reserved. No part of this publication may be
reproduced in any material form (including photocopying or
storing in any medium by electronic means and whether or
not transiently or incidentally to some other use of this
publication) without the written permission of the copyright
holder except in accordance with the provisions of the
Copyright, Designs and Patents Act 1988 or under the terms
of a licence issued by the Copyright Licensing Agency Ltd,
90 Tottenham Court Road, London, England W1P 9HE.
Applications for the copyright holder's written permission to
reproduce any part of this publication should be addressed
to the publishers

British Library Cataloguing in Publication Data
A catalogue record for this book is
available from the British Library

Library of Congress Cataloguing in Publication Data
A catalogue record for this book is
available from the Library of Congress

ISBN 0 7506 0556 1

LEEDS METROPOLITAN
UNIVERSITY LIBRARY
1700137268
5413
13.6.94
575.1 TEC

Composition by University of Greenwich
(formerly Thames Polytechnic)
Printed and Bound in Great Britain

The Biotol Project

The BIOTOL team

OPEN UNIVERSITEIT, THE NETHERLANDS
Prof M. C. E. van Dam-Mieras
Prof W. H. de Jeu
Prof J. de Vries

UNIVERSITY OF GREENWICH (FORMERLY THAMES POLYTECHNIC), UK
Prof B. R. Currell
Dr J. W. James
Dr C. K. Leach
Mr R. A. Patmore

This series of books has been developed through a collaboration between the Open universiteit of the Netherlands and University of Greenwich (formerly Thames Polytechnic) to provide a whole library of advanced level flexible learning materials including books, computer and video programmes. The series will be of particular value to those working in the chemical, pharmaceutical, health care, food and drinks, agriculture, and environmental, manufacturing and service industries. These industries will be increasingly faced with training problems as the use of biologically based techniques replaces or enhances chemical ones or indeed allows the development of products previously impossible.

The BIOTOL books may be studied privately, but specifically they provide a cost-effective major resource for in-house company training and are the basis for a wider range of courses (open, distance or traditional) from universities which, with practical and tutorial support, lead to recognised qualifications. There is a developing network of institutions throughout Europe to offer tutorial and practical support and courses based on BIOTOL both for those newly entering the field of biotechnology and for graduates looking for more advanced training. BIOTOL is for any one wishing to know about and use the principles and techniques of modern biotechnology whether they are technicians needing further education, new graduates wishing to extend their knowledge, mature staff faced with changing work or a new career, managers unfamiliar with the new technology or those returning to work after a career break.

Our learning texts, written in an informal and friendly style, embody the best characteristics of both open and distance learning to provide a flexible resource for individuals, training organisations, polytechnics and universities, and professional bodies. The content of each book has been carefully worked out between teachers and industry to lead students through a programme of work so that they may achieve clearly stated learning objectives. There are activities and exercises throughout the books, and self assessment questions that allow students to check their own progress and receive any necessary remedial help.

The books, within the series, are modular allowing students to select their own entry point depending on their knowledge and previous experience. These texts therefore remove the necessity for students to attend institution based lectures at specific times and places, bringing a new freedom to study their chosen subject at the time they need and a pace and place to suit them. This same freedom is highly beneficial to industry since staff can receive training without spending significant periods away from the workplace attending lectures and courses, and without altering work patterns.

Contributors

AUTHORS

M.R. Fowler, De Montfort University, Leicester, UK

Dr J.S. Gartland, Dundee Institute of Technology, Dundee, UK

Dr K.M.A Gartland, Dundee Institute of Technology, Dundee, UK

Dr P. Hooley, University of Wolverhampton, Wolverhampton, UK

Dr C. K. Leach, De Montfort University, Leicester, UK

Ir G. Tuijnenburg Muijs, Prins Bernhardkade, Rotterdam, The Netherlands

EDITOR

Dr K.M.A Gartland, Dundee Institute of Technology, Dundee, UK

SCIENTIFIC AND COURSE ADVISORS

Prof M. C. E. van Dam-Mieras, Open universiteit, Heerlen, The Netherlands

Dr C. K. Leach, De Montfort University, Leicester, UK

ACKNOWLEDGEMENTS

Grateful thanks are extended, not only to the authors, editors and course advisors, but to all those who have contributed to the development and production of this book. They include Ms H. Leather, Ms J. Skelton, and Professor R. Spier.

The development of this BIOTOL text has been funded by **COMETT, The European Community Action Programme for Education and Training for Technology**. Additional support was received from the Open universiteit of The Netherlands and by University of Greenwich (formerly Thames Polytechnic).

Contents

How to use an open learning text

An open learning text presents to you a very carefully thought out programme of study to achieve stated learning objectives, just as a lecturer does. Rather than just listening to a lecture once, and trying to make notes at the same time, you can with a BIOTOL text study it at your own pace, go back over bits you are unsure about and study wherever you choose. Of great importance are the self assessment questions (SAQs) which challenge your understanding and progress and the responses which provide some help if you have had difficulty. These SAQs are carefully thought out to check that you are indeed achieving the set objectives and therefore are a very important part of your study. Every so often in the text you will find the symbol Π , our open door to learning, which indicates an activity for you to do. You will probably find that this participation is a great help to learning so it is important not to skip it.

Whilst you can, as an open learner, study where and when you want, do try to find a place where you can work without disturbance. Most students aim to study a certain number of hours each day or each weekend. If you decide to study for several hours at once, take short breaks of five to ten minutes regularly as it helps to maintain a higher level of overall concentration.

Before you begin a detailed reading of the text, familiarise yourself with the general layout of the material. Have a look at the contents of the various chapters and flip through the pages to get a general impression of the way the subject is dealt with. Forget the old taboo of not writing in books. There is room for your comments, notes and answers; use it and make the book your own personal study record for future revision and reference.

At intervals you will find a summary and list of objectives. The summary will emphasise the important points covered by the material that you have read and the objectives will give you a check list of the things you should then be able to achieve. There are notes in the left hand margin, to help orientate you and emphasise new and important messages.

BIOTOL will be used by universities, polytechnics and colleges as well as industrial training organisations and professional bodies. The texts will form a basis for flexible courses of all types leading to certificates, diplomas and degrees often through credit accumulation and transfer arrangements. In future there will be additional resources available including videos and computer based training programmes.

Preface

The development of techniques for manipulating genes (genetic engineering) is based on an understanding of the molecular aspects of genetic processes occurring in nature. These techniques have, in turn, enabled expansion of our knowledge and understanding of how genes are organised and expressed in living cells and has provided major impetus to the development of biotechnological enterprises.

The techniques of gene manipulation enable us to transcend conventional biological boundaries to create completely novel gene combinations and even new 'designer' genes. The practical consequences of this are enormous and we are experiencing a phenomenal impact on many processes and products. New vaccines, therapeutic agents and diagnostic devices derived from this new technology are doing much to transform medical and veterinary practices; new enzymes and organisms are finding applications in food production and new crop varieties and crop protectants are being adopted in agriculture.

Although much has been achieved, it is widely believed that we are still at the beginning of realising the full potential that this technology has to offer. We can anticipate for the future that gene manipulation will have an increasing role to play in healthcare, food production, agriculture and, by enabling the introduction of cleaner technologies, in industry and environmental management.

The aim of this text is to provide readers with a thorough understanding of the techniques which underpin these developments. By gaining this understanding, readers will be able to apply the techniques of gene manipulation to tackle specific problems or to achieve desired objectives in a wide variety of fields.

The text predominantly centres on microbial systems which lie at the heart of many gene manipulation activities. A brief introductory chapter is provided to enable readers to orientate themselves within the BIOTOL series of texts dealing with genetic engineering. Then the natural genetic recombination systems which occur in micro-organisms and mutagenesis are reviewed since these are also harnessed in gene manipulation *in vitro*. Readers are then provided with an overview of the basic steps employed in gene manipulation to provide a context in which the rest of the text can be studied. The major part of the text examines these steps in greater depth dealing with the extraction of genetic material; gene isolation; the techniques that can be used to introduce DNA into cells; characterisation, modification and the design of vectors. A recurrent theme throughout is the need to develop appropriate strategies to identify the desired products of gene isolation and gene recombination. In the final part of the text, issues of safety concerned with using micro-organisms and genetically manipulated cells are discussed.

The techniques described in the text are becoming well established and, although they may become progressively refined, are likely to be used for many years and in a wide variety of systems. Although this text includes many specific examples, the applications of genetic manipulation to particular groups of organisms are described in a partner BIOTOL text, 'Strategies for Engineering Organisms'. The applications of gene manipulation in particular business sectors are described in the BIOTOL 'Innovations' texts. These are, 'Biotechnological Innovations in Health Care', 'Biotechnological Innovations in Animal Productivity', 'Biotechnological Innovations in Food

Processing', 'Biotechnological Innovations in Chemical Synthesis' and 'Biotechnological Innovations in Energy and Environmental Management'.

The author:editor team, composed of experienced teachers and respected researchers in this important area, have written a logically developed, highly interactive text which will ensure that readers will gain a firm understanding of this essential part of contemporary biotechnology.

Scientific and Course Advisors: Professor MCE van Dam-Mieras
Dr. CK Leach

Genetic engineering - the BIOTOL strategy

Genetic engineering - the BIOTOL strategy

1.1 Introduction

The growth of contemporary biotechnology can be attributed largely to the introduction and development of *in vitro* techniques for manipulating genetic information. These techniques not only enable us to rearrange existing genomes but also to create completely new mixtures of genes by transferring genes across interspecies boundaries and to even create novel genes. The application of these techniques has made important contributions to science and business. They have opened up new possibilities in a wide range of scientific endeavours and have led to increased understanding of how living organisms function. This new knowledge together with the application of genetic engineering techniques have had important consequences to health care, agriculture, food processing and in environmental and resource management. Genetic engineering must be regarded as an important enabling technology that provides great opportunities to exploit the large genetic resources available within the biosphere and to enlarge and enrich this resource.

This technology is, however, still in its infancy. Only relatively few of the enormous range of possibilities have, as yet, been realised. Even so, we can cite many important developments that have been dependent upon this new technology. These include: improvement in the yields of desirable products (eg antibiotics, hormones, enzymes); the synthesis of new, or hitherto inaccessible products (eg human proteins, vaccines); the development of crops with improved characteristics (eg disease resistance); new approaches to animal and crop production (eg micropropagation of ornamental plants, breeding programmes); new approaches to disease diagnosis and treatment; new approaches to quality control, forensic science (eg genetic fingerprinting) and to environmental issues (pollution monitoring and treatment). BIOTOL texts concerned with genetic engineering are designed to enable readers to acquire knowledge and skills in this centrally important technology thereby enabling them to contribute to a wide variety of endeavors and enterprises in which genetic engineering plays (or will play) a key role. The aim of this brief introductory chapter is to orientate the reader within the BIOTOL scheme and to explain the structure of the text.

1.2 BIOTOL texts and genetic engineering

The BIOTOL approach to genetic engineering is to divide the study into three major phases. The first, the theme of this text, is to explain how we might isolate, characterise and manipulate genetic material *in vitro*. The second phase, the topic of the BIOTOL text 'Strategies for Engineering Organisms', shows how such manipulated material may be used to generate organisms with particular desirable characteristics. This second BIOTOL text, builds on the knowledge provided by 'Techniques for Engineering Genes' by providing details of the genetic vectors available for manipulating particular groups of organisms. It covers the important groups of prokaryotes, fungi (especially yeasts), plants and animals. The third phase of the BIOTOL approach is to explain how these techniques may be applied in a variety of business sectors. These are explained in

BIOTOL'S 'Biotechnological Innovations' series and include; 'Biotechnological Innovations in Health Care', 'Biotechnological Innovations in Crop Improvement', 'Biotechnological Innovations in Animal Productivity', 'Biotechnological Innovations in Food Processing', 'Biotechnological Innovations in Chemical Synthesis' and 'Biotechnological Innovations in Energy and Environmental Management'. The overall BIOTOL scheme is shown in Figure 1.1.

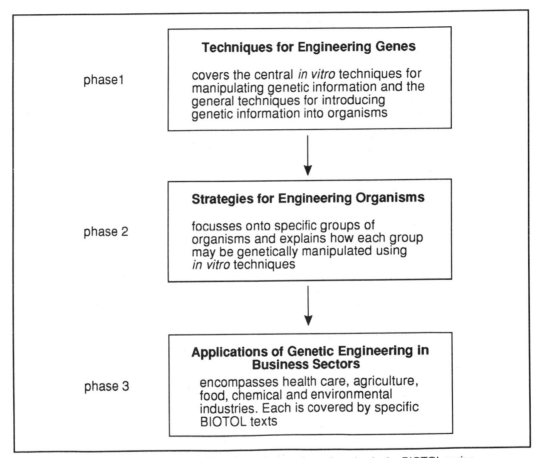

Figure 1.1 Phases in the development of knowledge in genetic engineering in the BIOTOL series.

1.3 Pre-requisite knowledge

Genetic engineering is an enabling technology underpinned by relevant biological sciences. The ability to manipulate genetic information implies the need to understand the nature (chemical and physical) of the genetic material and how it is organised in living systems. It also pre-supposes knowledge of how genetic information is transcribed and processed and how these processes are regulated. 'Techniques for Engineering Genes' has been written on the assumption that the reader is familiar with these aspects of the biological sciences. (Note that this essential pre-knowledge is covered in the BIOTOL texts, 'Genome Management in Prokaryotes' and 'Genome Management in Eukaryotes'). Despite this assumption, the authors have provided many helpful reminders embedded at appropriate places within this text.

1.4 Structure of the text

We begin with a chapter on *in vivo* techniques of genetic manipulation especially of micro-organisms. It is perhaps a little surprising to begin a text on what is essentially a discussion of *in vitro* techniques by considering *in vivo* techniques in micro-organisms. This is, however, of fundamental importance for a number of reasons. Firstly, the reader should recognise that the features of micro-organisms (fast growth rates, single-celled nature, wide metabolic capabilities, accessible genetic exchange system, ease of manipulation) makes these organisms particularly important in their own right and many biotechnological processes employ such organisms. Secondly, *in vivo* manipulations are often important as providers of vital information and are essential pre-requests to *in vitro* manipulation (eg by enabling us to map or modify gene arrangements and activities). Thirdly, even with the *in vitro* manipulation of genetic information from non-microbial systems, micro-organisms are used as hosts to produce desired genes or gene segments. Fourthly, although *in vitro* techniques open up possibilities not accessible through *in vivo* techniques, some desirable end points are possible by either route and thus in these instances we must regard the two groups of techniques as competitors. Finally, since micro-organisms are important both directly as biotechnological agents and indirectly by enabling manipulation of non-microbial genetic information, this chapter provides a reminder of the natural genetic exchange systems available amongst microbes.

In Chapter 3, we provide an overview of the basic steps of *in vitro* genetic manipulation. We describe, in outline, the core technique of nucleic acid extraction especially of bacterial DNA, plasmids and mRNA and then show how this genetic material may be cut into specific fragments, rejoined to other fragments and how the recombinant fragments may be identified. Subsequent chapters expand on this overview by examining particular aspects in greater detail. Chapter 4, for example, describes how DNA may be introduced into living systems. (This is developed more fully with specific groups of organisms in the BIOTOL text, 'Strategies for Engineering Organisms'). Chapter 5 explains how we may target our search for particular genes whilst Chapter 6 decribes how we may modify (mutate) such genes in a deliberate and predictable way. In Chapter 7 we explain how we may characterise particular genes. Here particular attention is placed on mapping and sequencing genes and on how we determine whether or not, and to what extent, specific genes are expressed. Chapter 8 explains the techniques used to screen genetic libraries and how regulatory sequences may be identified and manipulated. The final two chapters deal with issues of safety. Thus in Chapter 9, we consider the safety issues arising from the cultivation of micro-organisms and this is extended in Chapter 10 to the procedures that are adopted in handling genetically manipulated organisms. These latter two chapters are of particular importance since they alert the reader to the relevant regulatory issues involved in using such systems and provide guidance to appropriate procedures and behaviour within the laboratory.

In vivo techniques of genetic manipulation

In vivo techniques of genetic manipulation

2.1 Introduction

In this chapter, we are going to describe a range of techniques involving intact cells that may be used to genetically manipulate cells, especially bacteria. These techniques harness genetic processes that normally occur in nature. This chapter therefore has much in common with some of the material covered in the BIOTOL text, 'Genome Management in Prokaryotes', in which the molecular genetics and molecular biological processes of bacteria are described. The motivation for the inclusion of the material in this text and the emphasis placed on this material here is, however, somewhat different. Here we are predominantly concerned with hijacking naturally occurring genetic exchange processes to construct strains of bacteria of value to mankind either by providing new knowledge or by enabling the generation of strains with more desirable features for application in industrial processes. The harnessing of these naturally occurring processes is also an essential component of the profitable application of the products of manipulating DNA *in vitro*. Indeed, the successful application of the *in vivo* techniques of genetic manipulation be it genetic material from plant, animal or microbe, is entirely dependent upon the processes described in this chapter. It is therefore vitally important that you understand the material covered in this chapter if you are to properly appreciate the topics covered in later chapters.

In the first part of the chapter, we discuss the reasons why micro-organisms in general, and bacteria in particular, are objects of study and describe the major types of mutants that can be produced in these systems. The major part of the chapter, however, is devoted to the discussion of the three major groups of techniques that can be used to transfer genetic material in bacteria. These techniques are based on the process of transformation, conjugation and transduction. On completion of this chapter, you will have sufficient knowledge of these three groups of procedures that will enable you to make judgements concerning their potential to be used to transfer genes and to achieve genetic recombination.

In the final part of the chapter, we will briefly describe the importance of naturally occurring transposable genetic elements (transposons) as devices which enable us to modify gene activity and to identify and position (map) genes.

2.2 *In vivo* manipulation of micro-organisms

Much experience has been gathered over the years on the manipulation of micro-organisms *in vivo*. We can ask ourselves, therefore, why do we manipulate micro-organisms *in vivo*?

This question can be answered at three levels:

- much has been done (and still remains to be done) to understand important fundamental genetic, mutational, biochemical or regulatory processes. Many of the initial studies have achieved this using whole organisms;

- much work in microbial genetics involves constructing bacterial or yeast strains and viruses with particular combinations of genes for strain improvement purposes. Some of these targets can be achieved by *in vivo* manipulation;

- finally, even following the *in vitro* techniques discussed in Chapter 3, manipulated DNA must be re-introduced into a host, altered *in situ* or moved from one organism to another before use.

In vivo manipulations have resulted in improved strains being used in industrial processes such as in brewing, in chemical synthesis and in pollution control.

The micro-organisms covered in this chapter will be predominantly bacteria.

∏ Before reading on, see if you can list reasons for using bacteria.

We would anticipate that you would have included in your list of reasons the following:

- ease and simplicity of growth;

- rapid culture (short life cycle; 20-40 min);

- a viable bacterium can easily be detected as it can grow on a solid (agar gelled) nutrient medium to produce a colony. We can separate bacterial cells by spreading them onto a gelled medium. Each bacterium will divide clonally to produce a colony of many thousands of genetically identical individuals. This enables us to isolate and produce genetically pure strains;

phenotype

genotype

- selective techniques are commonly available to identify and select mutants (this is often a prerequisite for further studies). The ability of bacteria to grow and divide under various culture conditions is the major phenotypic (physical appearance in a particular environment) criterion used in bacterial studies. This contrasts with the term genotype which is used to describe the genetic make up of an organism.

2.3 Types of bacterial mutants and their selection

As we shall see later, it is important to have bacterial mutants to use in *in vivo* manipulations. (Note that the actual processes of mutagenesis are described in the BIOTOL text, 'Genome Management in Prokaryotes').

Mutants may be selected by:

- sensitivity to chemicals;

- requirement for certain compounds for growth;

- ability to use (breakdown) compounds.

2.3.1 Sensitivity to chemicals

sensitivity and resistance to antibiotics

Bacteria are sensitive to (unable to grow in the presence of) some chemicals. Often antibiotics such as streptomycin (Sm) are used. Sensitive organisms may be described as Sm S, resistant organisms as Sm R. Other antibiotics commonly used are penicillin (Pen) and its derivatives and kanamycin (Km). Obviously differences in sensitivity to antibodies provide us with an opportunity to select mutants.

2.3.2 Requirement for inclusion of specific nutrients in the culture medium

prototroph

autxotroph

Most bacteria may be grown on very simple (minimal) culture media in the absence of any organic substances other than a carbon source such as glucose or galactose. They are able to manufacture for themselves most complex molecules needed for growth such as amino acids, vitamins and lipids. Such an organism is said to be a prototroph. Some bacteria, however, have a dependence on specific nutrients being included in the culture medium, as they cannot make all essential compounds themselves. An auxotroph is an organism which requires the presence of an organic compound (not as carbon source) in the culture medium. For example, a strain requiring the presence of the amino acid proline (Pro) is a proline auxotroph, represented as pro^- and may be distinguished from the normal (or wild type) strain shown as pro^+. Other commonly used auxotrophic markers are leu^-, his^-, trp^- and thr^- (strains requiring; leucine; histidine, tryptophan and threonine respectively). These differences in nutritional demand may also be used as a basis for selecting mutants.

2.3.3 Ability to use substances as nutrient sources

Not only are some strains unable to make certain compounds, others are unable to utilise various compounds. For example the inability to use lactose or galactose is represented as lac^- or gal^- the wild type being lac^+ and gal^+. Such differences may also be used to select mutants.

SAQ 2.1

A typical minimal medium (MM) contains Na^+, K^+, Mg^{2+}, Ca^{2+}, Fe^{2+}, NH_4^+, Cl^-, phosphate buffered to neutrality (around pH 7), SO_4^{2-} and a carbon source, in this case galactose. Table 2.1 reports the ability of organism A to grow on such a minimal medium containing various supplements.

Having read the previous introductory paragraphs you should be able to confirm your understanding by ringing the answer which correctly describes organism A.

Organism A is a:

1) his^-, leu^- auxotroph, can breakdown galactose and is Sm S;

2) his^-, leu^- prototroph, can breakdown galactose and is Sm S;

3) his^+, leu^+ prototroph, can breakdown galactose and is Sm S;

4) his^+, leu^+ auxotroph, can breakdown galactose and is Sm S;

5) his^-, leu^- auxotroph, requires Sm for growth, can breakdown galactose.

Organism A	
Medium/supplement	Growth +/-
MM + galactose	-
MM + galactose + histidine	-
MM + galactose + histidine + leucine	+
MM + galactose + leucine	-
MM + galactose + histidine + leucine + streptomycin	-

Table 2.1 Growth of organism A in minimal medium (MM) with supplements.

From Table 2.2 , we can see organism B is a *pro⁻* auxotroph. Its inability to grow on MM + galactose unless glucose is added shows that it is unable to break down galactose.

Organism B	
Medium/supplement	Growth +/-
MM + galactose	-
MM + galactose + proline	-
MM + galactose + glucose	-
MM + galactose + proline + glucose	+

Table 2.2 Growth of organism B in minimal medium (MM) with supplements.

If we were to mix organism A described in SAQ 2.1 + organism B under the right circumstances we would find individuals capable of growth on MM + galactose. This outcome could be due to information transfer between the two strains of bacteria and is a common event in some bacterial species.

2.4 Transfer of genetic information in bacteria

Bacteria have several different ways of transferring genetic information, both to one another and to their direct progeny. These mechanisms may be divided into 3 main categories (which as we will see later can themselves be subdivided):

transformation: uptake of naked DNA into recipient bacteria;

conjugation: DNA is transferred from a donor (male) to a recipient (female) via a specialised sex pilus;

transduction: bacterial genes are carried from a donor to a recipient via a bacteriophage.

2.5 Bacterial transformation *in vivo*

Transformation is the process in which recipient cells take up naked DNA from the surrounding medium. Such DNA pieces can arise naturally (from dead, lysed bacteria), are generally large, on average 20 000 nucleotide pairs long, and might contain useful genes. Small pieces may also be taken up but a minimum length of about 500 basepairs is required in order for integration into the host chromosome and hence transmission to progeny cells. Many bacterial species are capable of such uptake eg *Bacillus subtilis*, *Escherichia coli*, *Haemophilus influenzae*, *Pseudomonas aeroginosa* and *Streptococcus pneumoniae*.

competent cells   Cells capable of DNA uptake are said to be competent. Naturally, perhaps 1 in every 100 cells is in such a state although, this number is variable as is the length of time that a cell may remain competent. The process is active (energy requiring). Bacterial transformation may also be artificially elicited in the laboratory (at higher frequencies using alternative techniques and DNA types). In addition it is possible to manipulate eukaryotic cells such as yeast, mammalian cultured cells and treated plant cells to obtain high frequency DNA uptake.

Transformation was once an important means of determining gene order in microbial genetics and remains still, for some organisms, the only means of mapping (ordering) genes.

In bacteria, the main genetic information (a few thousand genes) is present in the bacterial chromosome which is comprised of circular, double-stranded DNA in a highly condensed form (folded and coiled). This DNA is packed into the central region of the cell, it is not bounded by a nuclear membrane. This structure is known as a nucleoid.

The natural transformation process may be split into several stages.

It may be summarised by the following:

1) Reversible binding of double-stranded DNA molecules to cell surface receptor sites.

2) Irreversible uptake of donor DNA into those bacterial cells that are competent.

3) Conversion of donor DNA into single-stranded molecules via degradation of one strand.

4) Integration of all, or part, of the single-stranded donor DNA into the recipient chromosome.

5) Segregation and phenotypic expression of donor DNA in the recipient.

These stages are illustrated in Figure 2.1 .

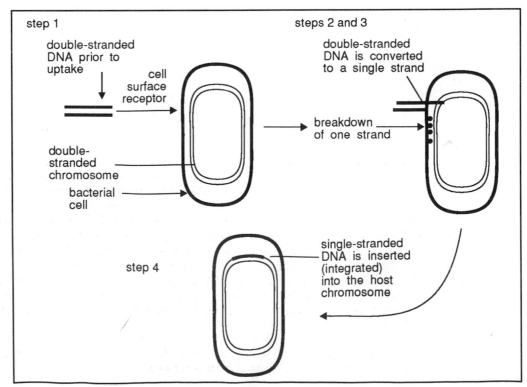

Figure 2.1 The natural transformation process. Note that we have shown the bacterial chromosome as a simple circle. In reality it is highly folded and compact.

Steps 1 to 3 occur even when DNA from diverse sources is used. However, for a high frequency of integration/recombination, homologous (similar) DNA is required. Integration of heterologous (foreign) DNA into the bacterial chromosome occurs more rarely.

Let us consider two genes *pro* and *trp*. The genotype of an organism able to make proline and tryptophan will be written *pro⁺*, *trp⁺*. Let us assume for the sake of argument that these genes are on opposite sides of the chromosome. When DNA is isolated from the bacterium, the chromosome will shear into smaller bits and the two genes will be on physically separate pieces of DNA (Figure 2.2).

We will assume that a very small amount of this DNA is mixed with a culture of living *pro⁻*, *trp⁻* (auxotrophic) bacteria of the same species, and that 1% of these bacteria are competent. As transformation is a relatively rare event it is unlikely that a particular recipient that is transformed for one gene will also be transformed for the other as this would require two transformation events.

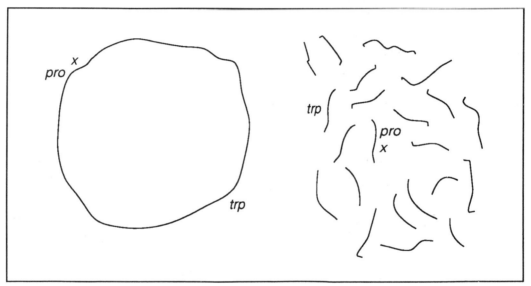

Figure 2.2 The shearing of DNA. Distant genes (eg *pro* and *trp*) are on separate DNA fragments. Closely related genes (*pro* and *x*) may be on the same DNA fragment (see text).

locus

single and double transformants

Let us consider in addition a gene, *x*, that is situated directly next to the *pro* gene position, or locus. It is highly likely that during DNA isolation these two genes will remain together on the same piece of DNA (see Figure 2.2). Therefore, there is a good possibility that if a recipient is transformed with one trait it will also be co-transformed with the second trait, associated with the adjacent genetic locus. Thus if two genetic markers are closely linked double transformants may be found at a frequency approaching the frequency of single marker transformants in a comparable single marker experiment.

So we can see that the frequency with which two markers are co-transformed can be used as a crude estimate of the distance between them.

SAQ 2.2

Which of the following experiments would you use to determine if y and z loci were close together on the bacterial chromosome.

Circle correct response.

1) Mix DNA from y^-, z^- bacteria with y^+, z^+ bacteria and plate onto a medium lacking both y and z.

2) Mix DNA from y^+, z^+ bacteria with y^-, z^- bacteria and plate onto a medium lacking both y and z.

3) Mix DNA from y^+, z^+ bacteria with y^-, z^- bacteria and plate onto three media one lacking y another lacking z and the third lacking both.

An experiment shows that the following genes are co-transformed:

1) A and B

2) E and D

3) F and E

4) D, B and C.

∏   What is the probable gene order? (Write down your answer before moving on).

Their sequence can be found from this information, because in order for two genes to transform together they must be close together (assuming only very small amounts of DNA are used). Therefore we can tell that A and B must be near neighbours as are E and D and F and E. E must be between D and F as F and D do not co-transform. C must be between B and D because it co-transforms with D and B but not with A, E or F.

The order therefore is A B C D E F (or F E D C B A).

Information on the relative gene position is very important in strain construction. It is not always easy to select for the presence of a gene of interest and in such a situation by knowing which genes are near neighbours, we can initially look for the presence of a suitable selectable marker gene instead and assume that it is likely its neighbour is present too.

For instance, in the previous example, if gene B's product made host cells resistant to streptomycin (Sm R) and gene D's product made cells resistant to kanamycin (Km R) but we had no way to ensure the presence of the nearby gene C (which might for instance be one of several genes effecting yield of an industrially important product), we could look for cells which were Sm R and Km R and assume that they would possess gene C as well.

Many bacteria naturally lyse and release DNA on ageing or in various deleterious conditions. This production of homologous DNA in a bacterial population, together with the process of transformation, has probably been important during evolution.

2.6 Plasmids and conjugation

2.6.1 Plasmids

Up until now we have considered the bacterial chromosome as the main source of genetic information. In addition to the bacterial chromosome, bacteria often also contain one or more genetic elements called plasmids.

General properties of plasmids

supercoiled DNA

Plasmids are double-stranded circular supercoiled DNA molecules and are relatively small (0.2 - 4% of the size of the bacterial chromosome). They do, however, contain important genes which, although dispensable by the host in most growth conditions, have particular value in certain environments. For example, conferring on the host the ability to grow in the presence of streptomycin which is only of benefit to the host organism in the presence of such an antibiotic but has little or no effect on the bacterium in the normal environment.

replicon

A plasmid is a replicon (unit of genetic material capable of independent replication) that is stably inherited in an extra-chromosomal state. In the following section, we will concern ourselves with the many plasmids of *E. coli*. This organism, found in the human gut, is a commonly chosen model for genetic studies.

E. coli plasmids can be split into three main groups: F, R and Col types, the properties of which are summarised in Table 2.3 and are discussed in the following sections.

Plasmid	Type	Conjugative self transmissible	Information
F	sex fertility plasmid/ episome	all	mediate self transfer from cells with plasmid (F⁺ donor/male) to cells without, (F⁻ recipient/ female): can cause chromosome transfer (see Hfr later)
R	drug resistance plasmid	most	makes host resistant to one or more antibiotics - may transfer to cells lacking it
Col	colicinogenic plasmid	some	code for proteins called colicins that can kill closely related bacteria that lack a Col plasmid of the same type

Table 2.3 A summary of the common *E. coli* plasmid types.

episome

We need to distinguish the terms 'plasmid' and 'episome'. A plasmid exists in an autonomous state, in which it replicates independently from the chromosome. In contrast, an episome exists in an integrated state, in which it is inserted into the host chromosome.

R-factors are of particular relevance to mankind. They are readily transferred, even between species, eg from *E. coli* to *Shigella* and *Salmonella*. Pathogenic bacteria such as these with multiple antibiotic resistances now constitute a serious and ever increasing problem in both human and veterinary medicine.

The F (fertility) factor is of particular importance in *E. coli* as we shall find out.

It can exist in two different states:

- as a plasmid (autonomous state);

- as an episome (integrated state).

A donor cell with an F factor in an autonomous state is described as F⁺. Upon conjugation (see Section 2.6.2). with a recipient bacterium (F⁻) only the F factor is transferred and both exconjugants (cells involved in conjugation) become F⁺. Transfer can occur once or twice per generation. Thus, by mixing a population of F⁺ cells with F⁻ cells, virtually all cells in the new population quickly become F⁺. Such rapid transfer is, however, not the case for most transmissible plasmids. Only about 0.2% of a population of cells containing R plasmids are competent donors (which is just as well for us). Most R plasmids encode a repressor which acts on genes required for plasmid transfer. Thus following transfer of the plasmid into a recipient, which lacks the repressor, further transfers may occur almost immediately. But, once the repressor is made, fertility decreases. This phenomenon is known as fertility inhibition. It should be noted that fertility inhibition is not a property of all R plasmids. Interestingly, the R repressor may also reduce the transfer of the F factor.

fertility inhibition

Incompatibility groups

Although many plasmids use the host's machinery to replicate themselves, there are many significant differences in the mechanisms of plasmid replication. Each type of plasmid possesses its own genes for controlling the rate of replication initiation and hence the copy number per cell. We may classify plasmids on this basis. Stringent control - low copy number plasmids exist at only 1 or 2 copies per cell. The partitioning of such plasmids at cell division is generally controlled to avoid total plasmid loss from the cell.

low copy number plasmids

Relaxed plasmids - may be present at high copy number (10-100 per cell). Thus if a cell is transformed with one copy of a high copy number plasmid it will replicate rapidly to reach a high number. This process is controlled by a plasmid encoded repressor that negatively regulates initiation of replication, ie - the more plasmid present the more repressor present which limits its further replication. Repressor activity may depend on concentration. So, when a cell enlarges before dividing into two, the repressor concentration drops and replication occurs until twice the number of plasmids are present, at which time sufficient repressor is present to prevent further replication.

high copy number plasmids

For high copy number plasmids, a greater concentration of repressor is necessary to limit replication than for low copy number plasmids.

Let us pose a question. Can more than one type of plasmid co-exist in the same host cell?

To answer this question, we will describe two situations. In the first, consider the presence of two plasmids, one is a relaxed plasmid the other a plasmid under stringent control. What happens if the replication of these two plasmids was controlled by the same repressor? Under these conditions, we would expect to lose the low copy number (stringently controlled) plasmid as the culture grows. The production of the repressor as a consequence of the presence of the relaxed plasmid would switch off the production of the low copy number plasmid. The average number of these low copy number plasmids per cell would, therefore, progressively fall as the cells grew and divided. As a consequence of this, many progeny cells would therefore fail to receive a copy of such plasmids. The co-existence of these two types of plasmids in the same host cell is unlikely. They are said to be incompatible. Hundreds of plasmids have been sorted out into incompatibility (Inc) groups. We can test for incompatibility by determining whether or not different plasmids can cohabit in a cell. Plasmids from the same Inc group cannot co-exist.

In the second example, we can visualise the situation in which the two types of plasmids are controlled by different repressors. Clearly the regulation of plasmid numbers are independent of each other and, in such circumstances, the plasmids can co-exist. Experiments have shown that many (up to 7 different types) of such compatible plasmids may co-exist in a single cell.

SAQ 2.3	Consider a cell with one copy of plasmid (A) and one copy of plasmid (B) both under stringent replication control. We can designate such a cell as (A,B). What is the likely plasmid content of cells after fifty cell divisions if the plasmids have the same repressor?

Plasmids with degradative/virulence functions

degradative
plasmids

There are two other plasmid categories worthy of note at this stage, before we pass to a discussion of conjugation. The first category are the degradative plasmids found in bacteria such as the genus *Pseudomonas*. These plasmids allow the host to metabolise unusual substances such as toluene and salicylic acid. Examples of such plasmids are the TOL plasmids of *Pseudomonas putida*.

virulence
plasmids

Virulence plasmids come into a second category and confer pathogenicity on the bacterial host. The Ti and Ri plasmids of *Agrobacterium* are well known examples of these and are respectively responsible for the induction of crown gall and hairy root diseases in plants.

2.6.2 Conjugation

Conjugation is the process of transfer of DNA from a donor to a recipient cell via a specialised intercellular connection or conjugation tube, known as a pilus (the plural is pili). This process does not occur naturally in many species, although it is common in *E. coli, P. aeroginosa* and *Vibriocomma*.

This form of genetic transfer is very different to syngamy in eukaryotes. Syngamy is the fusion of two cells resulting in an equal genetic contribution from each parent to the progeny in a cross. Conjugation, however, is a one way transfer and no reciprocal exchange of information takes place. It is therefore a valuable means of genetic manipulations *in vivo* as genes of importance may be moved from one host to another.

Conjugation, DNA transfer and replication

We may summarise the conjugation process as follows, using *E. coli* as an example:

- formation of donor/recipient pairs following effective cell/cell contact.

This is achieved by the use of specialised sex pili - these are specialised cell surface hairlike tubes called F or R pili depending on whether their synthesis is controlled by an F or R plasmid. They bring about initial contact and then draw cells into close contact.

conjugative
and
non-conjugative
plasmids
drivers/mobilis-
ing plasmids
Conjugative (transmissible) plasmids may provide conjugation facilities for other non-conjugative plasmids which would not on their own mediate DNA transfer. They may be used experimentally to allow movement of an otherwise non-transmissible plasmid from one host to another. Such plasmids are known as drivers or mobilising plasmids.

- preparation of DNA for transfer (mobilisation).

Some plasmids are able to prepare their own DNA for transfer. In this process, a plasmid encoded protein makes a single-stranded nick in the transfer origin (*ori* T) of the plasmid. Such mobilisation is usually plasmid specific. Some depend upon other plasmids to prepare their DNA for transfer.

- DNA transfer

The 5′ end of the nicked strand is then transferred through the pilus into the recipient this is coupled to a form of replication called rolling circle replication in which the displaced strand is replicated in the recipient and the circular strand is replicated in the donor. The process is illustrated schematically in Figure 2.3. Examine this figure carefully as it provides an overview of conjugation and DNA transfer.

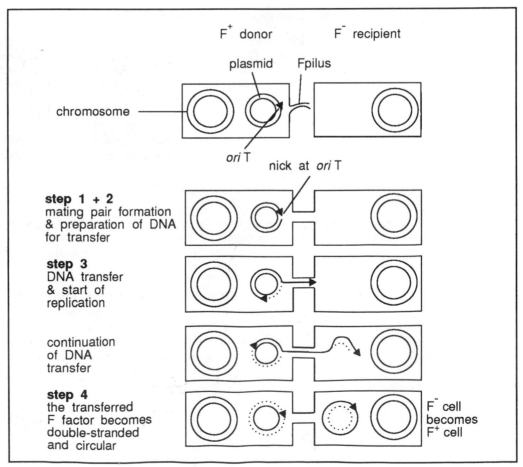

Figure 2.3 The F⁺/F⁻ conjugation process.

F mediated gene transfer

Hfr cells

An F factor can integrate into the host chromosome at one of many sites and in either orientation. Such integration is thought to be mediated by insertion (IS) elements (these will be described later). A cell which contains an integrated F factor is known as Hfr, which stands for high frequency recombination - the reason for this will soon become apparent.

Hfr conjugation

In such a state, the F factor still mediates its self-transfer as we have seen previously, however this time, the attached chromosome may also follow into a recipient F⁻ cell (see Figure 2.4).

a)   The F factor (represented as the dark circle, with the four reference points a, b, c and d) integrates into the bacterial chromosome (the white circle with the numbers 1 - 6 representing different parts of the chromosome) to form an Hfr.
Note the arrow head represents the first part of the DNA to be transferred during conjugation, see b) below.

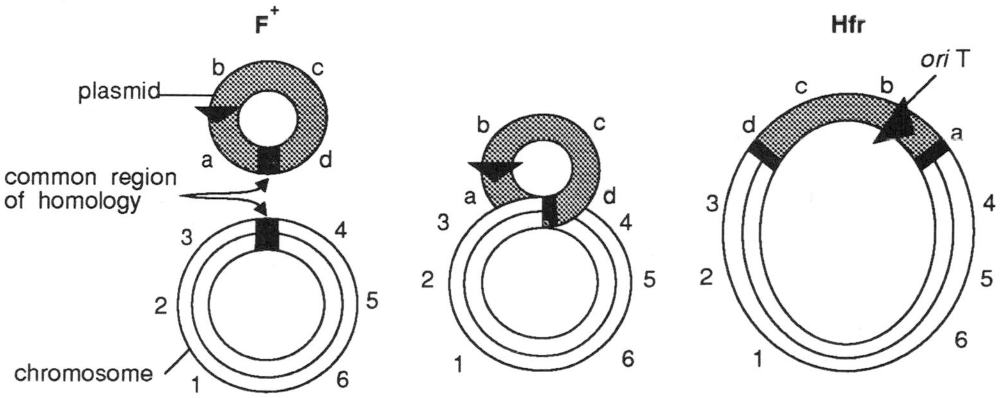

b)   One strand of Hfr begins to transfer into the F⁻ cell. Replication occurs via the rolling circle mechanism as in Figure 2.3.

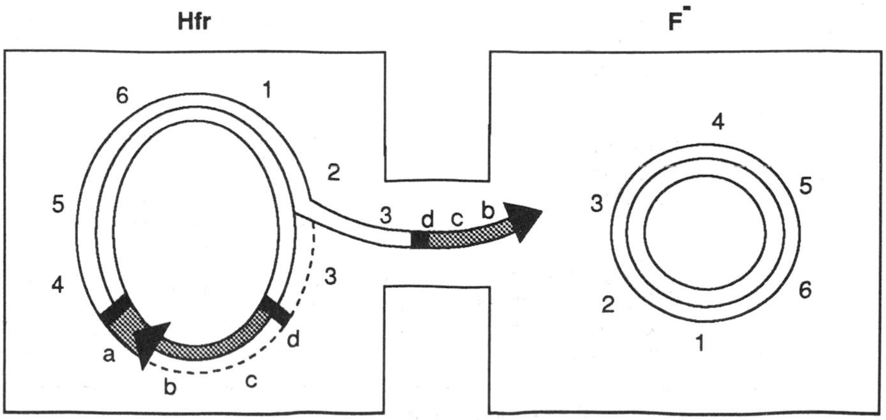

Figure 2.4 Hfr conjugation.

Only very rarely will the whole chromosome pass through the delicate conjugation tubes as these often break and the cells separate before the entire transfer process is complete. As a consequence of this, in Hfr x F⁻ matings, the recipient will usually remain F⁻ as a portion of the integrated F factor (marked a) in Figure 2.4) is the last to be transferred. Only occasionally will the entire F factor be successfully transferred in which case the recipient becomes Hfr. This is in contrast to F⁺ x F⁻ matings where the F⁻ becomes F⁺ (Figure 2.3).

Chromosome transfer proceeds at a fairly steady rate and takes about 90-100 minutes to complete, this is in contrast to the 2-3 minutes taken by F factor transfer. In the laboratory the mating can be stopped after different time intervals by violent agitation such as that produced by a blender, after which the extent of transfer can be determined.

In the same way that our discussion on transformation showed us that markers which are close to each other are likely to be transferred together into a recipient, two markers that are close together on the chromosome are more likely to be transferred and integrated together following recombination into the recipient chromosome.

As shown in the example below for the visualisation of recombinants, appropriate selection measures must be available to select against both parental types .

∏ We will now describe a typical experiment.

A Hfr *his⁺, leu⁺, trp⁺, pro⁺*, SmS donor was mixed with an F⁻, *his⁻, leu⁻, trp⁻, pro⁻*, SmR recipient at time zero.

After the periods specified in Table 2.5 aliquots were removed and conjugation tubes disrupted using a Waring blender. Samples were plated on minimal media (MM) with additives (see Table 2.4 for media codes).

Medium	Code
MM	MM
MM+his,leu,trp,Sm (no pro)	-p
MM+leu,trp,pro,Sm (no his)	-h
MM+his,trp,pro,Sm (no leu)	-l
MM+his,leu,pro,Sm (no trp)	-t

Table 2.4 Media codes.

Table 2.5 gives the number of bacterial colonies found on each type of plate (MM, -p, -h, -l or -t). Conjugation was interrupted after the times specified. The volume sampled was the same in each case.

Medium	Number of colonies																
	Time (minutes) of conjugation																
	0	2.5	5	7.5	10	12.5	15	20	25	30	35	40	45	50	55	60
MM	100	100	100	100	100	100	100	100	100	100	100	100	100	100	100	100
-p	0	0	0	6	12	19	26	30	30	31	30	30	29	29	29	29
-h	0	0	0	0	0	0	0	0	0	0	0	0	1	7	8	8
-l	0	8	19	30	40	40	40	41	40	40	39	38	39	38	37	37
-t	0	0	0	0	0	0	0	0	0	7	15	19	19	18	18	18

Table 2.5 Results of the experiment described in the text. The results record the number of colonies produced after plating out cells onto various media after the cells had been allowed to conjugate for varying lengths of time.

Sm prevents the Hfr from growing. The absence of any of the auxotrophically required amino acids (his, leu, trp or pro) will prevent growth of the F⁻ recipient unless it receives genes encoding the capacity to synthesise these from the donor.

We can visualise this process in the following way:

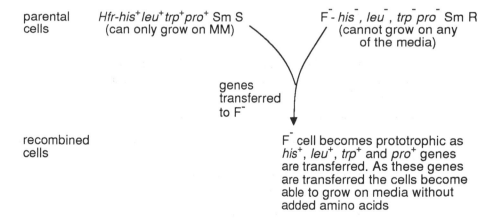

From the minimal medium, we can work out the total number of Hfr cells present in each sample as the Hfr can grow and the F⁻ cannot. From the other media we can work out the % Hfr cells that have donated DNA including markers (*his, leu, trp,* or *pro*) to F⁻ cells. Only such recombinants are capable of growth on -p, -h, -l and -t which lack one of these amino acids but contain Sm.

Plot the data for each of the makers on a graph of % recombined cells for each marker against conjugation time.

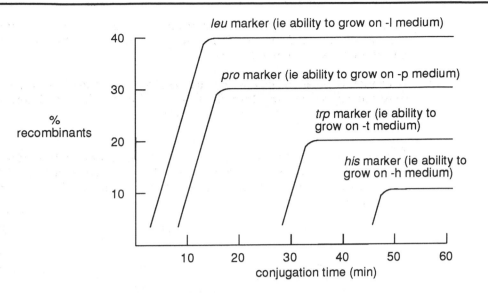

You should see from your graph that there is a time before which no recombinants are detected. Each curve has a linear region which may be extrapolated back to the time axis and may be used to give a value for the time at which each marker first enters the bacterium. This should give you the times for the first appearance of the markers as:

leu - 1.5 min

pro - 6 min

trp - 26 min

his - 44 min

In fact the time taken for genes to appear is a good indication of the distance the genes are from the origin of transfer. A map distance of 1 minute corresponds to the length of chromosome transferred in 1 minute during conjugation.

We can use this information to construct a map of the chromosome (see Figure 2.5).

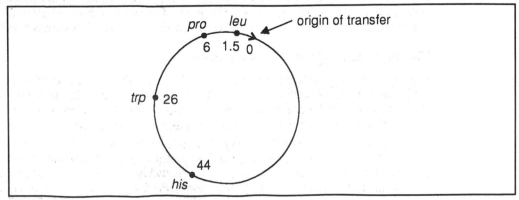

Figure 2.5 Chromosome map constructed from the interrupted mating experiment described in the text.

The number of recombinants for each type reaches a plateau value, which is lower the later the marker enters the recipient (this is mainly due to the fragility of the conjugation tube). Transfer begins at a particular point and genes are transferred in order either clockwise or anticlockwise depending on the orientation of the F factor in the chromosome as shown in Figure 2.5.

Note that the F factor may be inserted in a variety of positions in the genome and thus different Hfr strains will transfer genes at different times in interrupted conjugation experiments.

∏ Consider the three Hfr strains below. Note the orientation of the origin of transfer. 1) What will be the order in which genes are transferred? 2) Will the time difference between the transfer of genes a and b be the same or different in each case?

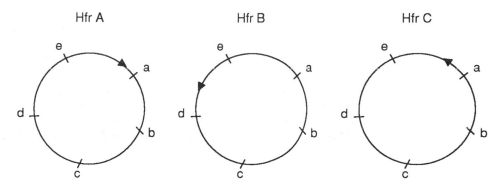

1) HfrA will transfer genes in the order e d c b a;

HfrB will transfer genes in the order e a b c d;

HfrC will transfer genes in the order a b c d e.

2) The time difference between a and b being transferred will be the same in each case. Although the genes are being transferred at different actual times, the distance between a and b is the same in each case.

Using such a variety of strains it is possible to construct a chromosome map. One produced for *E. coli* is shown in Figure 2.6. The map is divided into minutes and, by convention begins with 0 at the *thr* A (threonine) locus and continues to 90 or 100.

The importance of recombination in conjugation

In the preceding sections we have seen how DNA is transferred into cells. It must undergo a process called recombination in order for it to integrate into the host chromosome and be maintained through subsequent cell divisions. In the absence of recombination, we would not be able to detect recombinant colonies following the type of mating (Hfr x F⁻) we discussed previously in this section. This is due to the fact that in bacteria, with the exception of a few phage, only circular DNA which contains a suitable replication origin is able to replicate and hence pass to future generations. This, we should note, is also true of DNA taken into the cell during transformation. In fact, linear DNA is degraded by nucleases. In F⁻ x F⁺ matings the situation is obviously different as it is likely the entire plasmid will be transferred and will therefore be able to maintain itself and replicate in its circular form.

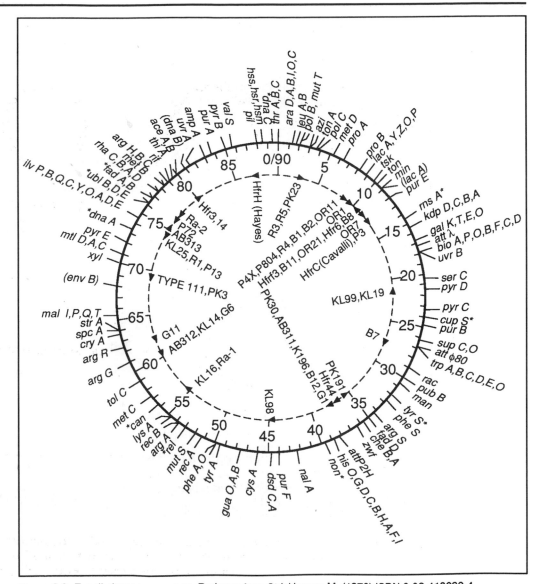

Figure 2.6 *E. coli* chromosome map. Redrawn from Strickberger M. (1976) ISBN 0-02-418090-4, MacMillan Publishing Co Inc, New York.

In Hfr and F⁻ matings or following transformation with linear DNA, it is necessary to have two (or an even number) of recombination events flanking the gene(s) to be incorporated in order for the incoming linear DNA to be integrated into the circular chromosome (see Figure 2.7 a). With only one (or an odd number) of recombination events (see Figure 2.7 b), it would loose its circular nature and be unable to replicate properly.

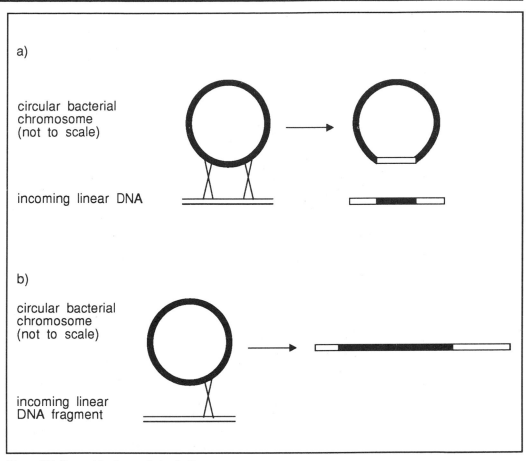

Figure 2.7 Recombination and circularity.

Time of entry experiments such as those previously discussed give an overall picture of the *E. coli* genome but cannot give us much fine detail.

The recombination frequency in *E. coli* is quite high and recombinations may occur several times between markers as little as a few minutes apart. Thus the linkage mapping scheme explained in the transformation section (Section 2.5) is not applicable to such large DNA segments.

In order to explain this consider the following. Using time of entry experiments we have demonstrated a gene order:

leu, pro, trp, his at 1.5, 6, 26 and 44 minutes respectively.

In order to show close linkage we would assess colonies showing recombinant properties for one prototrophic marker gene (for example *pro*⁺) for the presence of the others.

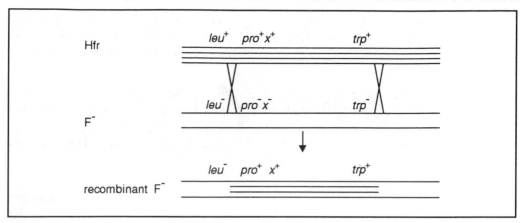

Figure 2.8 Two recombination events leading to the co-transfer of *pro⁺ x trp⁺*.

If just two recombination events occurred as above in Figure 2.8, then the alleles *pro⁺* (6.0 minutes), x (6.05 minutes) and *trp⁺* (26 minutes) would be integrated into the F⁻ chromosome. However in reality further recombination events would occur, for example:

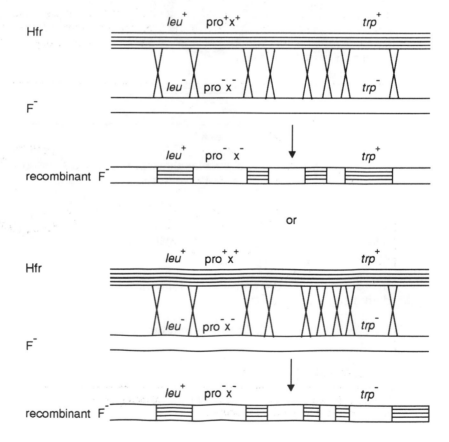

As the distance between two genes increases the probability of a recombination event between them increases as well. Thus only genes very close together like *pro* and *x* are unlikely to be subject to such events and are likely to stay linked together in the recipient.

At this stage it is important to mention that recombination events are controlled by a variety of enzymes. Not all bacteria possess a full set of these enzymes. Recombination enzyme-deficient bacteria are known as *rec*⁻ and are not used during strain construction processes that require recombination events. These enzymes are discussed in more detail in Chapter 3.

F mediated sexduction

As well as integrating into the bacterial chromosome to become an Hfr, it is possible, although more rare, for the F factor to be excised (cut out). Sometimes the excision process is faulty and produces an F factor which contains bacterial genes. The number of these genes excised varies from 1 to half of the chromosome. These altered F factors F prime (F') are called F prime (F').

Transfer of F' into a recipient occurs in F'/F⁻ matings. This process is called sexduction.

As the F' factor replicates autonomously the F⁻ becomes F', and establishes a new line of F', cells. The resulting line will be partially diploid, if, as is likely, it carries its own copy of the region of the bacterial chromosome carried by the F', element (see Figure 2.9). This system is itself very useful in the study of dominant/recessive relationships at genetic loci.

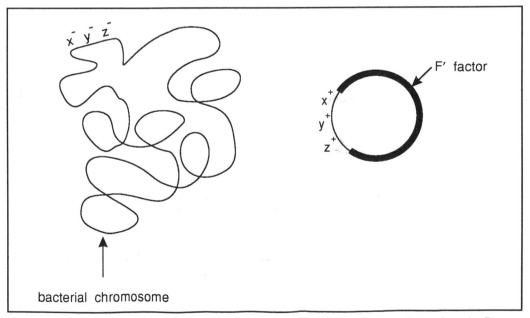

Figure 2.9 F' cells are partially diploid (meridiploids) for x y z as these genes are carried both on the F' factor and on the bacterial chromosome.

You are provided with a Sm S Hfr culture of bacteria of the type described in Figure 2.5, but are told that in addition there is an arg^+ allele at 79 minutes.

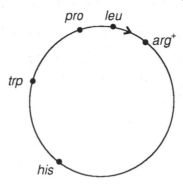

This strain is mated with a $rec^-, leu^-, pro^-, trp^-, his^-, arg^-$, SmR, F$^-$ recipient culture. Five colonies result after selection on minimal medium with Leu, Pro, Trp, His and streptomycin but lacking Arg.

Which of the following may be true? Ring the true answers.

1) The 5 colonies are arg^+ prototophs.

2) The Hfr has conjugated with the F$^-$ and transferred the arg^+ gene which has integrated into the recipient chromosome.

3) The Hfr strain contained some F′ plasmids. An F′ plasmid containing the arg^+ allele has been transferred.

As we can see it is possible to manipulate strains and move desired features around from one bacterium to another. In the case of sexduction only genes close to the site of insertion of the F factor will be transferred. Other more flexible methods also exist.

2.7 Transduction

2.7.1 The basics of transduction

We are now going to look at an alternative process for transferring DNA. This process is called transduction and involves the use of viruses that infect bacteria. Such viruses are known as bacteriophage (phage). Transduction is the process by which bacterial DNA is transferred by phage particles from a donor bacterium to a recipient cell.

We can discriminate between two types of transducing phage:

• generalised transducing phage;

• specialised transducing phage.

generalised transducing phage

Generalised transducing phage produce particles that contain only DNA obtained from the host bacterium rather than phage DNA. This DNA may be derived from any part of the bacterial chromosome.

specialised transducing phage

Specialised transducing phage produce particles that contain both phage and bacterial DNA linked together to form a single molecule. The bacterial genes are obtained from a particular region of the bacterial chromosome.

The transduction mechanism depends on how the phage grows in bacterial cells.

There are two alternative life cycles by which phage reproduce:

- the lysogenic cycle (involving a stage known as lysogeny);
- the lytic life cycle.

In order to understand the process of transduction we will briefly consider each life cycle in turn with the commonly used phage lambda (λ phage) as an example. Figure 2.10 shows the structure of phage lambda.

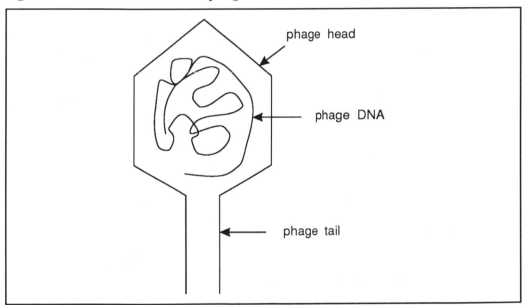

Figure 2.10 Structure of phage lambda (highly stylised).

2.7.2 Lambda lysogenic cycle

In lysogeny, the phage DNA enters the bacterium and, rather than being replicated rapidly, it inserts into the bacterial chromosome (in this state the phage is known as a prophage). As the bacterial cell divides, the phage DNA is then replicated and passes, along with the bacterial chromosome, to the daughter cells.

In its integrated state, the phage lambda lytic genes are repressed (turned off). A lysogenic cell is immune to secondary infections by the same virus, as the lytic genes of the incoming virus would also be repressed.

This situation can be explained in the following way:

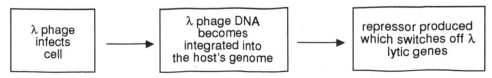

If a lytic λ phage now enters this cell, it will fail to cause lysis because of the repressor produced as a result of the integration of the lysogenic phage into the bacterial chromosome. We shall describe the lytic cycle in the next section.

temperate phage Phage that possess both lytic and lysogenic life cycles are described as temperate. Temperate phage are capable of transduction. Examples of such phage include lambda and phi80.

The chromosome of such temperate phage integrates at only one, or a few, specific attachment site(s) within the host chromosome.

Transition from the lysogenic to the lytic state occurs naturally (at a rate of about 1 out of 10^5 phage) or may be induced by, for example, irradiation with ultra violet light.

2.7.3 Lambda lytic cycle

The prophage may excise and enter the lytic cycle. On commencing the lytic cycle the prophage is excised from the chromosome and begins to replicate autonomously. Progeny phage are assembled and are released from the bacterial cell when it lyses. We can see the result of this process. If healthy bacteria are grown on an agar plate they **plaque formation** form an opaque layer of growing cells. Where bacterial lysis and phage release occurs a clear area, known as a plaque, is formed. Both lytic and lysogenic pathways are illustrated in Figure 2.11

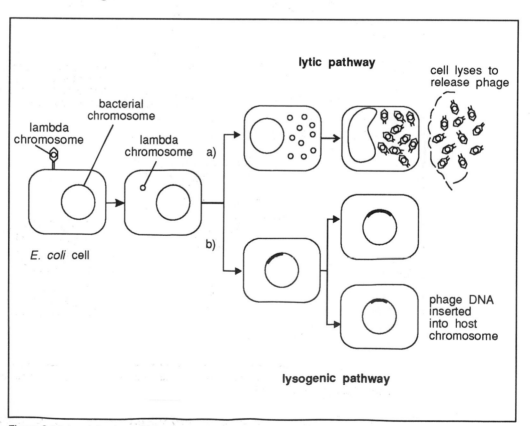

Figure 2.11 Lambda phage life cycles.

Like the integration process, the excision process is site specific since both processes involve phage enzymes.

2.7.4 Specialised transduction

The excision process of temperate phage is generally precise, however, occasionally (1 in 10^6 excision events) a mistake is made. Host genes close to the site of insertion may be excised with phage DNA and packaged into phage particles. This will occur only if the spacing between the 2 cut sites gives rise to a molecule 76-106% of the length of the normal lambda genome. An example of this is given in Figure 2.12.

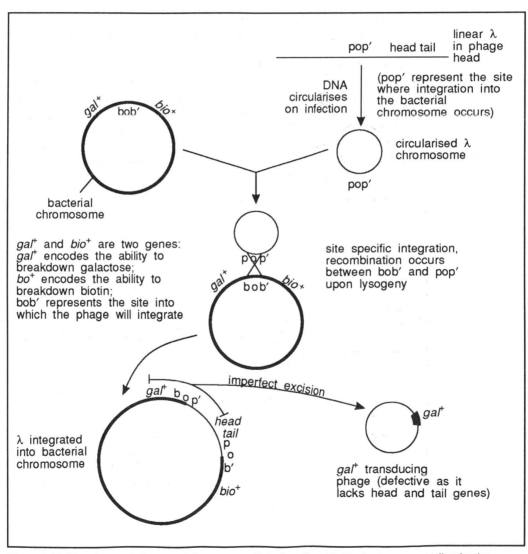

Figure 2.12 The abnormal phage excision process. Note 'head' and 'tail' refer to genes coding for the construction of phage heads and tails.

The number of missing genes depends on the position of the cuts. Genes encoding head and tail proteins are essential, so for example the *gal*⁺ transducing particles illustrated in Figure 2.12 are defective and cannot form plaques.

2.7.5 Generalised transduction

virulent phage

In contrast to the temperate phage we have just read about, virulent phage always multiply and lyse the host cell after infection.

Generalised transduction is mediated by some virulent bacteriophage and by certain temperate bacteriophage whose chromosomes do not integrate at a specific site of attachment on the host chromosome (for example phage Pl). In the latter case generalised transducing particles are only produced during the lytic stage.

Generalised transducing particles contain only bacterial DNA. This DNA may be derived from any part of the bacterial chromosome. On entering the lytic cycle the phage proteins are synthesised and new phage assembled by packaging phage DNA in heads. The packaging mechanism however, is not always accurate, and sometimes (with a frequency of around 1 in 10^6) will package a piece of bacterial chromosome instead yielding a phage body containing only bacterial DNA.

complete
transduction

This fragment of bacterial DNA may enter a new cell and become integrated into the host chromosome after pairing with its homologous region and subsequent recombination. In this state, the integrated DNA will be inherited and transduction is known as complete.

abortive
transduction

Alternatively, it may remain free and not integrated. Under these circumstances it will not be replicated. Hence it will remain in only one of the two daughter cells, leading to only one cell of a colony harbouring the fragment. This is known as abortive transduction.

Identification of transduced cells employs selective techniques similar to those discussed previously.

2.7.6 Use of transduction for genetic maps

Generalised transduction is often used for mapping studies. About 1% of the host chromosome is usually picked up. Thus only a few genes that are in the phage will be introduced into the recipient cells, and only genes that are close together on the host chromosome will undergo co-transduction. If the transduction frequency is 1 in 10^5, we would expect a single cell to be transduced twice at a frequency of 1 in 10^{10}, ie very rarely.

Specialised transduction introduces even smaller pieces of chromosomal DNA and allows fine scale mapping or genetic manipulation.

SAQ 2.5	If, for example, we wanted to transduce a *trp⁻ E. coli* strain to *trp⁺* we could isolate P1 phage following their infection of *trp⁺* bacteria and use these phage to infect *trp⁻* bacteria.

SAQ 2.5

If, for example, we wanted to transduce a trp^- *E. coli* strain to trp^+ we could isolate P1 phage following their infection of trp^+ bacteria and use these phage to infect trp^- bacteria.

When performing P1 phage transductions only one particle in 10^5 carries bacterial DNA. This means that a large number of active phage are present. It is therefore desirable to prevent these phage from infecting and lysing (or lysogenising) the transductants.

Given the fact that bacteriophage P1 requires Ca^{2+} to attach to *E. coli* and that Ca^{2+} can be removed from the medium by adding sodium citrate which binds it, which of the following would permit maximum recovery of transduced *E. coli*?

1) Mix phage with excess bacteria. Before bacteria lyse to produce new phage remove all Ca^{2+}. Plate on selective medium lacking Trp but containing sodium citrate.

2) Mix excess phage with a small number of bacteria. Before the bacteria lyse to produce new phage remove all Ca^{2+}. Plate on selective medium lacking Trp but containing sodium citrate.

3) Mix phage with excess bacteria. Plate on selective medium lacking Trp but containing Ca^{2+}.

4) Phage mixed with bacteria which are already infected with a mutant P1 phage which remains lysogenic and will not enter the lytic phase. Plate on selective medium lacking Trp.

2.8 Transposons

A large portion of our study up to now has assumed that genes are in fixed positions and do not move sites. However some DNA sequences can and do change position. These mobile DNA sequences are called transposable genetic elements or transposons and are typically relatively small (500-100 000 bases long). Such sequences are found in a diverse range of organisms such as bacteria, fungi, insects, plants and mammals. Bacterial transposons are in the lowest size range (often < 1500 bases long) and often contain only genes involved in promoting or regulating transposition.

transposons

2.8.1 Insertion sequences

insertion
sequences

Insertion sequences (IS) are examples of such simple transposons. Strains of *E. coli* possess at least 5 such elements predictably named IS1, IS2, IS3, IS4 and IS5.

These mobile sequences move (transpose) once every 10^4-10^7 generations and contain only the few genes involved in transposition.

Insertion sequences are present on many pieces of DNA. The example we have seen so far involved the Hfr. Here insertion elements on both the bacterial chromosome and the F plasmid provide regions of homologous DNA which allows recombinational insertion of the F plasmid into the bacterial chromosome in the formation of the Hfr. The different Hfr strains that occur are due to such F factor insertions at different sites and in different orientations. Different bacterial strains harbour insertion sequences in variable positions. The nucleotide sequences at each end of the insertion sequence are inverted repeats of each other. This arrangement is shown in Figure 2.13.

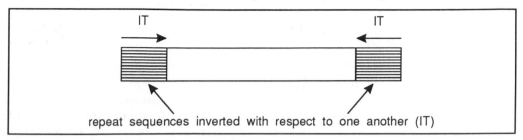

Figure 2.13 Structure of an insertion sequence.

2.8.2 Composite transposons

composite
transposons

Sometimes two adjacent homologous IS elements combine with other genes to form what is known as a composite transposon designated as Tn. This nomenclature also includes transposons that do not include IS elements but do contain at least one gene in addition to those needed for transposition. Such composite transposons can transpose themselves by the action of the flanking elements. Examples of such elements are Tn5 and Tn9. In Figure 2.14 we have illustrated Tn9 which carries a chloramphenicol resistance gene. Note that it is composed of two IS elements bordering the chloramphenicol resistance gene.

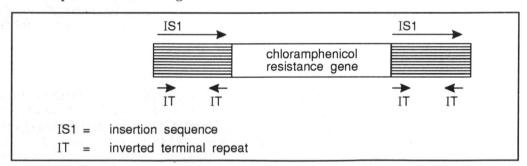

Figure 2.14 Structure of a composite transposon.

Bacterial transposons are capable of carrying many antibiotic resistances around, for example:

Tn 9 chloramphenicol

Tn 5 kanamycin

Tn 10 tetracycline

Tn 3 ampicillin

This is thought to have been important in the rapid distribution of the R plasmids which we discussed previously.

2.8.3 Transposon induced mutagenesis

Transposons do not only insert themselves between genes but also within them. Such insertions have the effect of inactivating the gene. This can be illustrated in the following way:

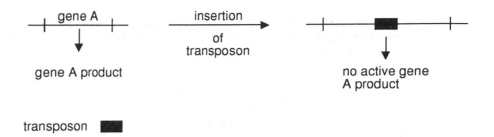

Genes that have been mutated by insertion of a transposable element may be said to have been tagged and assuming the transposon involved possessed a selectable marker such as an antibiotic resistance, the gene can be readily identified. We can illustrate this in the following way:

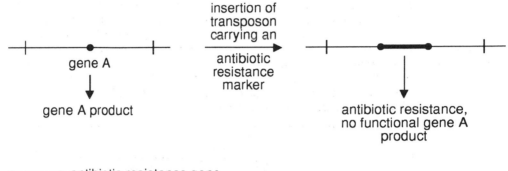

The result of such a transposon insertion is the co-occurrence of antibiotic resistance and the loss of gene A product. This, when used together with the gene cloning and hybridisation techniques discussed later in this book, provide a very useful method for identifying gene sequences in mixtures of DNA.

Transposable elements are also responsible for chromosome breakage and rearrangements. Under a variety of circumstances, transposons can cause deletion of genes or parts of genes. They may also cause segments of DNA to invert.

In vitro mutagenesis using transposons is an important technique in genetic engineering and will be described in detail later in this text.

At this stage we should note that transposons are not capable of moving from one bacterial cell to another as an infectious agent unless they are inserted into other pieces of DNA which may be transferred to a cell by transformation, conjugation or transduction.

SAQ 2.6

Ring the phenomena which may be mediated by transposons.

1) deletion;

2) pili production;

3) transposition;

4) mutation;

5) DNA inversions;

6) breakage and rearrangement of DNA;

7) movement of antibiotic resistance genes;

8) generalised transduction.

2.9 Bacterial protoplast fusion - an alternative approach to strain improvement

So far in this chapter we have concentrated on *E. coli* for our discussions of transformation, conjugation and transduction. Mechanisms for such events are well studied in this organism. Unfortunately, many industrially important micro-organisms lack a well studied mating system and are less amenable to transformation and transduction. Although protoplast fusion cannot be truly regarded as an *in vivo* method of genetic manipulation, we mention it here because it involves genetic recombination involving intact protoplasm.

Protoplast fusion has been developed to breakdown the barriers to genetic exchange in conventional systems. Bacterial cells are surrounded by a cell wall. Protoplasts (cells lacking walls) may be formed by treating whole cells with a variety of lytic enzymes. An osmotic stabiliser has to be included to prevent cell lysis as the cells no longer have a rigid cell wall to protect them. The fusion process usually involves polyethylene glycol (PEG) although some types of protoplast may be fused following the brief application of an electric field, given appropriate conditions. Protoplast fusion is now a routine technique for the *Streptomyces* species. The fusion of *S. griseus* and *S. tenjimariensis* has resulted in the production of the indolizine antibiotic, indolizomycin.

2.10 Concluding remarks

In this chapter we have described a variety of ways in which genetic exchange can be achieved within bacterial systems. We have also explained how transposable genetic elements can change the activities of specific genes. Thus, these processes enable us to copy, transpose, rearrange and mutate genes and to bring about new combinations of genes. These *in vivo* processes, therefore, offer a wide range of opportunities to modify the genotype and, therefore, the phenotypes of these micro-organisms. These opportunities have been grasped in a wide range of academic and commercial enterprises to provide new knowledge and to generate new strains of organisms for use in industrial processes.

There are, however, limitations to the targets that can be achieved if we are restricted to simply using these naturally occurring processes. For example, although human DNA may be taken up by the transformation mechanism by bacteria, this DNA is unlikely to be integrated and expressed within its new host cells.

In other words being dependent solely on the natural processes of genetic exchange imposes major restrictions on the combinations of genes that can be achieved. The advent of *in vitro* techniques such as restriction enzyme technology has changed this significantly. Using these new techniques coupled to the use of the mechanisms described in this chapter, we are able to achieve gene combinations that had hitherto been impossible and to modify the structure of individual genes in a refined and controlled way.

In the following chapters we will describe the *in vitro* techniques for manipulating DNA and show how these techniques together with harnessing transformation, conjugation, transduction and transposons have enabled us to generate organisms with more desirable characteristics.

Summary and objectives

In this chapter, we have described the major mechanisms of genetic exchange that occur in bacteria and explained how these mechanisms are of importance to both *in vivo* and *in vitro* genetic manipulation. We focused attention on to the processes of transformation, conjugation and transduction. In each case, we described the processes involved and then discussed the level of genetic exchange that can be achieved by these processes. We also explained how these techniques may be used to map and identify genes. In the final part of the chapter, we described the occurrence of transposons and their potential in modifying gene activity and in gene identification.

Now that you have completed this chapter you should be able to:

- understand the terms prototroph, auxotroph, wild-type, plasmid, episome;

- appreciate different types of bacterial mutants and understand their uses in *in vivo* manipulations;

- contrast F, R, Col, virulence and degradative plasmids;

- understand how to use cotransformation experiments to determine gene order;

- explain the role of the F factor in conjugation between F^+/F^-, Hfr/F and F'/F^- bacteria;

- explain how plasmid copy number may be controlled and what incompatibility groups are;

- describe the differences between generalised and specialised transduction and the differences in the life styles of the phage that bring about these events;

- understand why recombination and integration are necessary for DNA introduced into a bacterial cell to be expressed and passed to future generations;

- explain why protoplast fusion is the chosen means of manipulating some organisms;

- list the phenomena which may be mediated by transposons;

- appreciate the limitations of *in vivo* manipulations.

An introduction to *in vitro* genetic manipulation

An introduction to *in vitro* genetic manipulation

3.1 Introduction

In Chapter 2, we learnt that it is possible to manipulate genes in bacteria using the naturally occurring mechanisms of gene exchange. These manipulations have enabled us to learn much about the organisation and functioning of genes in such systems and have also allowed the production of strains which have improved characteristics for industrial purposes. However, some important limitations exist. For example it is difficult to envisage how these mechanisms by themselves can be employed to transfer desirable genes between entirely unrelated organisms. Furthermore, it is also difficult to see how we can use these mechanisms to deliberately alter existing genes in a controlled way to produce genes whose products have desirable new characteristics. The advent of techniques which enable us to modify DNA *in vitro* has however changed this perspective. The recombinant DNA technologies, usually referred to as genetic engineering, open up possibilities of transferring genes between quite unrelated organisms and to 'fine tune' the structure of individual genes so that they encode products with modified properties.

This chapter is designed to provide you with an overview of the basic steps of gene manipulation using the techniques of *in vitro* recombinant DNA technology. Thus we will examine the techniques for extracting DNA (and RNA) from a variety of sources and discuss the strategies used for purifying these molecules.

We will then go on to describe how these molecules may be fragmented in a controlled way and explain how we can separate and identify the fragments of interest. We then will describe how these pieces can be inserted into vehicles (vectors) which enable us to introduce these DNA fragments into host cells. We will include a brief description of these vectors and explain how we can use them to clone (make multiple copies of) the desired fragments. In other words, this chapter provides a discussion of the core techniques used in genetic engineering. Subsequent chapters will build on this discussion by examining aspects of genetic engineering in greater detail.

This is quite a long chapter so do not attempt to cover it all in one sitting. It is important that you fully understand the material covered in order to gain full advantage of the remaining chapters.

3.2 Recombinant DNA technology

∏   What do we mean by recombinant DNA technology?

Let us pause at this stage and think about it. For instance which of the following terms do you recognise as being associated with genetic engineering? Ring those you think are.

1) cloning;   2) vector;

3) restriction enzyme;   4) ligase;

5) agarose gel electrophoresis;   6) transformation;

7) plasmid.

In fact all of these terms/processes may be used at a certain stage of the process.

Examine Figure 3.1 carefully as it gives a summary of the steps used in *in vitro* recombinant DNA technology. There may be some terms used on this figure which are unfamiliar to you. Do not be worried by them at this stage; by the time you have completed this chapter, you will be able to fully understand the details described in this figure. It would be a sensible idea to copy this figure and to refer to it as you work your way through the chapter. In this way, you will have an easy reference point to where you are in the general scheme of *in vitro* recombinant DNA technology.

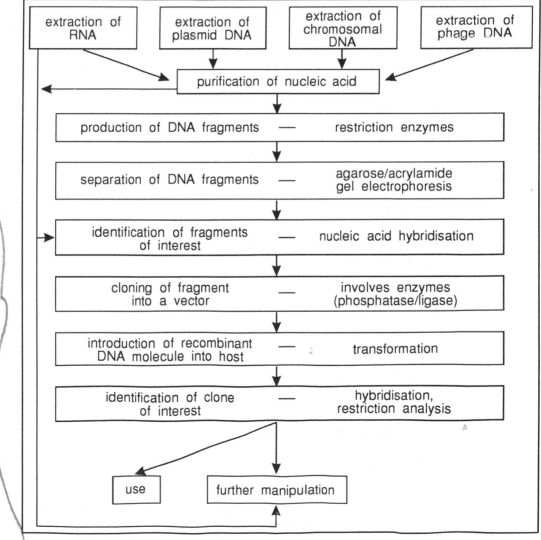

Figure 3.1 Basic steps in genetic manipulation.

As we can see from Figure 3.1 the first step in any manipulation process is the isolation and purification of the DNA we intend to manipulate. This DNA will include both the gene(s) of interest and the vector which will carry it. The vector is important for, as we shall see, it allows propagation and use of the gene of interest.

3.3 Nucleic acid extraction

3.3.1 Sources of nucleic acid

In order to perform *in vitro* genetic manipulation techniques, the genetic engineer may need to isolate several types of nucleic acid:

- the source from which each gene to be manipulated and cloned is to be obtained - this is likely to be total cellular DNA from plant/animal tissue or from a bacterial/yeast culture;

- plasmid DNA, often used as the intermediate or final cloning vector for relatively small pieces of DNA;

- phage DNA, used as a vector for cloning large DNA fragments;

- RNA, which may be needed in the cloning/screening process.

3.3.2 Methods for nucleic acid extraction

The process of nucleic acid preparation may be summarised in six stages.

- growth/cultivation of the starting material;

- harvesting of suitable tissues;

- breaking the cells open to release the nucleic acid;

- removal of components other than nucleic acid;

- purification of a particular type of nucleic acid (total cellular DNA, plasmid DNA, phage DNA or RNA);

- concentration of the isolated nucleic acid.

In the following sections (3.4-3.7) we will examine how DNA and RNA may be extracted from a variety of sources.

3.4 Preparation of total DNA

3.4.1 Total bacterial DNA

Growth and harvesting

We will not examine this aspect in detail here as it employs standard techniques. Bacterial cultures are readily grown in a variety of media and separation of bacterial cells from the culture medium is achieved via centrifugation to yield a pellet of concentrated cells. Usually the cells are washed in buffer and resuspended in a suitable medium to produce a thick paste.

Breaking open the bacterial cell wall

chemical lysis

lysozyme

SDS

DNA can be released from bacterial cells by chemical lysis. The type of chemical used depends on the bacterial species involved. With the most commonly used bacterium, *E. coli*, the enzyme lysozyme is used to digest the rigid cell wall structure. EDTA is added to remove the magnesium ions which are essential to preserve the structure of the cell envelope. A detergent such as SDS is included to solubilise the phospholipids in the cell membrane, thereby disrupting it and releasing the cell contents (see Figure 3.2 stages a and b).

Removal of components other than DNA

RNase treatment

phenol extraction

Centrifugation removes insoluble cell debris (stages c and d in Figure 3.2) such as cell wall pieces and membrane fragments. RNA may be degraded using the enzyme RNase (stages d and e Figure 3.2). Proteins can be removed by a variety of methods. We can use proteases to hydrolyse the proteins to produce water-soluble amino acids and peptides. Alternatively, the aqueous suspension containing DNA, proteins and the products of RNA hydrolysis may be shaken with phenol or phenol and chloroform. The aqueous phenol emulsion can then be separated by centrifugation. The proteins (which contain both hydrophobic and hydrophilic amino acid residues) collect at the interphase. The water-soluble DNA and RNA components remain in the aqueous phase (stages e and f in Figure 3.2).

Concentration of DNA

precipitation of DNA

In the presence of monovalent cations (eg Na^+), the addition of ethanol causes polymeric nucleic acids to precipitate. They may then be collected by centrifugation (stage g in Figure 3.2). Alternatively, if the DNA is sufficiently concentrated then it may be 'hooked' out of solution by winding it onto a glass rod. The ribonucleotides (small breakdown products of RNA) are too small to be precipitated and will be left in the aqueous-ethanol phase.

Usually Na^+ is added to a concentration of about 0.3 mol l^{-1} and ethanol to about 2 - 2.5 volumes of the aqueous phase. You should realise, however, that the scheme presented here is a generalised one. Different organisms require slightly different methods to achieve lysis and many workers have particular preferences. For example, some workers prefer to use elevated temperatures during treatment with SDS to denature enzymes (particularly DNase) in the bacteria. Obviously the temperature must not be too high because this would also denature the DNA.

Often the DNA collected at stage g) is re-dissolved in buffer and the phenol extraction and ethanol precipitation stages are repeated to remove further traces of protein. The collected DNA may be re-dissolved and dialysed against buffer to remove the traces of low molecular weight components (especially hydrolysed RNA components). Care has also to be taken so that DNases are not introduced during the purification procedure. These enzymes can come from the fingers of the researcher or from centrifuge tubes and from dialysis tubing. The latter is normally boiled before use to inactivate such enzymes. At this stage, however, we intend to simply alert you to these aspects rather than provide you with a lot of specific details; these are best considered as you gain experience in the laboratory.

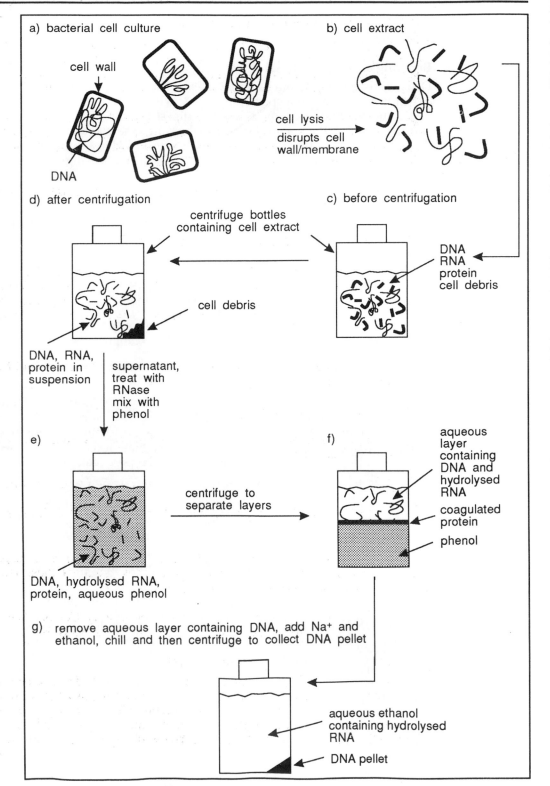

Figure 3.2 Preparation of total bacterial DNA.

LEEDS METROPOLITAN UNIVERSITY LIBRARY

3.4.2 Total animal/plant/yeast cell DNA

Obviously bacteria are not the only source of DNA
programmes. Other sources of total DNA such as plant, an
used. The same basic DNA purification steps may be used
with bacterial cells, although the means of breaking the cel

Physical means of breaking cell walls are often applied to
tissue in liquid nitrogen and grinding with a mortar and pes
which lack a cell wall, are often easily lysed by detergent treatment. Yeast cell walls may
be either disrupted enzymatically or mechanically.

The subsequent purification steps are similar to those described for bacterial cells.

3.5 Preparation of bacterial plasmid DNA

Preparation of this type of DNA differs from total DNA in that it has to be separated
from the bulk of the unwanted chromosomal DNA. This can be difficult, but is essential
if we are going to undertake cloning operations. When considering this, we must first of
all decide what are the characteristic differences between plasmid and chromosomal
DNA.

∏   Can you think of the characteristics of chromosomal and plasmid DNA enabling
    their separation?

The sort of characteristics we hoped you could recall were:

• plasmids are less than 10% of the size of the bacterial chromosome, and generally a
  lot smaller than this;

• very large pieces of DNA are likely to break into smaller pieces if handled roughly ;

• it should be possible to separate linear fragments of DNA from small, circular
  plasmids.

These are the features of plasmids which we can use readily to separate plasmid DNA
from chromosomal DNA. Watch out for the employment of these features in our
description of the methods used to purify plasmid DNA.

Several alternative means for the purification of plasmid DNA exist. We will consider
two of these. They both commence with a gentle means of cell lysis.

3.5.1 Controlled cell lysis/cleared lysate

controlled cell
lysis

The controlled cell lysis step is common to both methods and is achieved by including
sucrose with the lysozyme and EDTA step. This results in cells which although partially
wall-free do not burst straight away (see Figure 3.3). They may later be burst in a
controlled manner by the addition of a non-ionic detergent such as Triton X-100.
Plasmid DNA is released from the cell together with large pieces of chromosomal DNA.

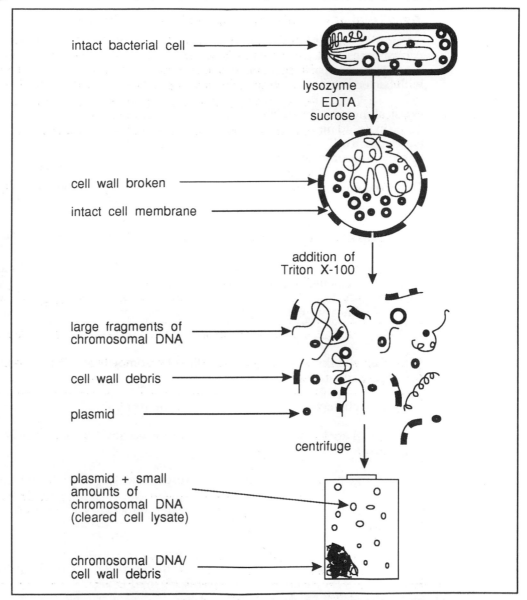

intact bacterial cell

lysozyme
EDTA
sucrose

cell wall broken

intact cell membrane

addition of
Triton X-100

large fragments of
chromosomal DNA

cell wall debris

plasmid

centrifuge

plasmid + small
amounts of
chromosomal DNA
(cleared cell lysate)

chromosomal DNA/
cell wall debris

Figure 3.3 Cleared lysis. The cell wall is partially broken with lysozyme and EDTA. In the presence of sucrose the cell does not burst. Careful addition of the detergent Triton X-100 causes gentle cell lysis producing a mixture of intact plasmids, large fragments of chromosomal DNA and cell wall debris and proteins. Gentle centrifugation removes the large fragments of chromosomal DNA and cell wall debris leaving intact plasmids and small amounts of chromosomal DNA in suspension.

cleared lysate

Centrifugation of this solution will remove cellular debris and some of the chromosomal DNA, giving what is known as a 'cleared lysate'.

It is from now on that the two types of method diverge, but both rely on differences between plasmid and chromosomal DNA. In order to understand these methods, we will need to remind you of some of the properties of plasmids.

3.5.2 Plasmid conformation as a key to purification

CCC
plasmid DNA

OC
plasmid DNA

Plasmid molecules in the bacterial host are double-stranded, circular entities. If both strands are intact, the molecule is described as covalently closed circular (CCC) DNA. The appearance of such a molecule is shown in Figure 3.4 a) below. The twisted nature of the molecule is due to the additional supercoils imposed on the structure.

If one strand is broken (nicked) the plasmid takes an open circular form (OC) see Figure 3.4 b).

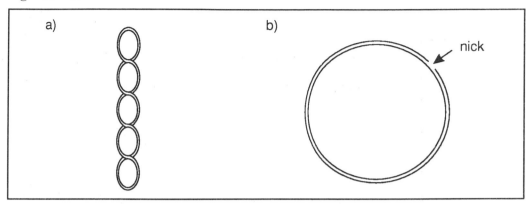

Figure 3.4 a) CCC plasmid DNA also sometimes referred to as supercoiled (SC) DNA; b) OC plasmid DNA.

If two nicks are present the DNA becomes linear, see Figure 3.5. (This is the same form as the pieces of broken chromosome).

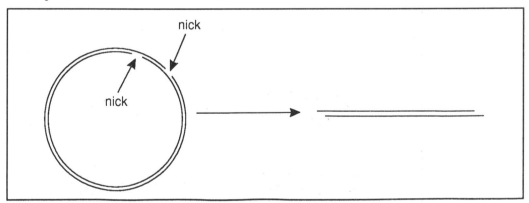

Figure 3.5 A plasmid with two nicks becomes linear.

This information is used in the two methods described schematically in Figure 3.6 (method 1) and Figure 3.7 (method 2).

3.5.3 Plasmid isolation (method 1)

In this method, use is made of the fact that linear double-stranded DNA is denatured by being exposed to high pH values in the range pH12-12.5 whilst supercoiled DNA is resistant to these conditions. Thus, at pH12-12.5, hydrogen bonds between the strands of DNA in the linear molecules are broken to produce linear single-stranded products. When the pH is lowered to 7 usually by adding potassium acetate, hydrogen bonds can be reformed between strands. If this is done at a cool temperature then mismatched strands can form stable hybrids to produce a large insoluble clump (aggregate).

The aggregated DNA can be removed from the supercoiled (covalently closed circular) DNA by centrifugation. These stages are illustrated in Figure 3.6.

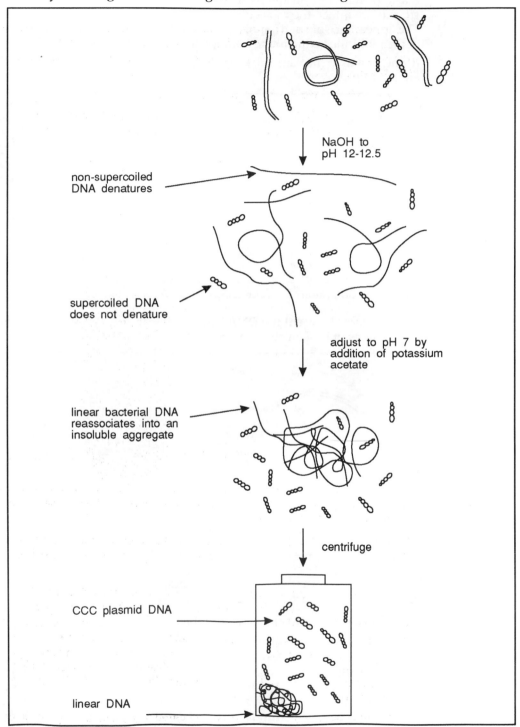

Figure 3.6 The selective preparation of covalently closed circular DNA (plasmid DNA) using an alkaline denaturation step (see text for details).

The supernatant is then collected and the plasmid DNA can be precipitated by the addition of Na⁺/ethanol at low temperature. The precipitated plasmid DNA is collected by centrifugation.

This method has the advantage that the treatment of the cell lysate with NaOH (see Figure 3.6) and subsequently adjusting the pH to 7 also results in the denaturation of proteins and RNA which also become insoluble and can be removed with the chromosomal DNA.

3.5.4 Plasmid isolation (method 2)

If a solution of caesium chloride (CsCl) is spun at high speed, centrifugal forces set up a density gradient in the tube with dense CsCl solution at the bottom and a less dense solution at the top.

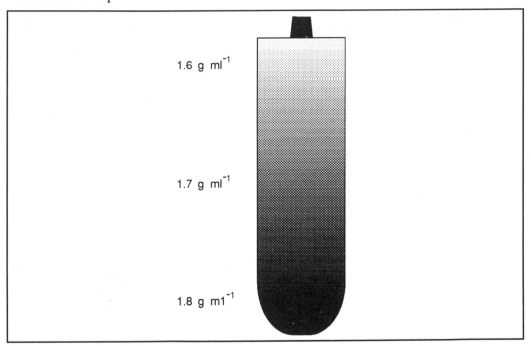

Figure 3.7 CsCl gradient formation. Note the figures on the left-hand side of the tube record the density of the CsCl solution in the centrifuge tube.

Molecules present in this solution will float at their buoyant density. Thus, in practice, DNA will float in the central zone at about 1.7 g ml⁻¹, RNA will pellet at the bottom and protein will float at the top.

We can include ethidium bromide in the solution. Ethidium bromide is a planar molecule which enables it to slot into the DNA helix causing it to unwind to some extent. The CCC DNA, having no free ends, has only a limited ability to unwind. Both the linear and OC DNA by virtue of their free ends can unwind to a large extent and their buoyant densities decrease and they form a band in a different position to the CCC DNA (see Figure 3.8). The position of the DNA bands can be seen clearly when the tube is irradiated with UV light, as DNA/ethidium bromide fluoresces with an intense orange colour under such conditions.

CCC DNA may subsequently be removed by puncturing the tube with a needle attached to a syringe. The ethidium bromide can then be removed by partitioning with butanol and the plasmid DNA precipitated with Na⁺/ethanol at low temperature.

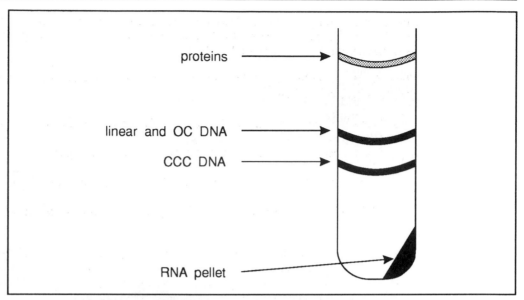

Figure 3.8 A CsCl density gradient following centrifugation. Note that ethidium bromide is added to the cleared lysate preparation. This unwinds linear and OC DNA making these molecules less dense. CCC DNA only unwinds to a very limited extent and is denser than linear and OC DNA under these conditions. Note the relative positions of RNA, CCC DNA, linear and OC DNA and proteins (see text for details).

3.6 Bacteriophage DNA isolation

In order to obtain high yields of bacteriophage DNA it is essential to grow the bacterial host in such a manner as to achieve maximum numbers of phage particles. Unlike methods to purify plasmid or total cellular DNA, the phage can generally be obtained without artificial cell breakage, although this is sometimes required. Note that the phage particles are released during the phage lytic cycle (see Chapter 2).

Intact cells and the remains of any lysed cells can be removed from the infected culture medium by centrifugation. Phages are then collected from the clarified supernatant by precipitation (this is often achieved using polyethylene glycol) and further centrifugation. Protein extraction using phenol or a protease removes phage coat proteins. Phage DNA can then be further purified by CsCl density gradient centrifugation as described in Section 3.5.4.

3.7 Isolation and purification of RNA

3.7.1 Total RNA

importance of
RNase and
RNase
inhibitors

RNA is very sensitive to breakdown by RNases. In order to obtain good preparations of RNA, it is essential to minimise the activity of the RNases which are liberated during cell lysis. This is achieved by including chemicals which inactivate RNases as soon as they are released. All glass and plasticware and all chemical solutions involved in experimentation must also be RNase free. This may be achieved, for example, by treating with dry heat (180°C for 2 hours) or by using a chemical RNase inhibitor such as diethylpyrocarbonate solution, followed by autoclaving. RNase activity can be

diethyl-
pyrocarbonate

minimised by keeping cells and their extracts cold. All procedures are performed where possible at 4°C and tissues/cells that are harvested for RNA preparations are either used immediately or else stored at -70°C.

The extraction method used depends on the source of the RNA, but in general follows the steps outlined below:

- cells are broken to release nucleic acid in the presence of RNase inhibitors;

- cell wall and membrane debris are removed by centrifugation;

- contaminating proteins are removed, often using phenol and chloroform;

- nucleic acid is collected following precipitation with Na^+/ethanol at low temperature;

- contaminating DNA is removed by centrifugation on a 5.7 mol l^{-1} CsCl cushion (the RNA will go to the bottom of the tube, the DNA will not penetrate the CsCl) or by repeated washing and resuspension in 3 mol l^{-1} sodium acetate (the DNA is soluble in this concentration of sodium acetate, the RNA is not).

3.7.2 Isolation of poly (A$^+$) RNA

The vast majority of mRNAs from animal, plant and yeast cells carry tracts of poly (A) at their 3' termini. Such mRNA can therefore be separated from the bulk of cellular RNA by affinity chromatography using a solid support (which can be small cellulose beads or specially treated paper) to which is attached a polythymidine residue.

The poly (A) terminus of the mRNA will bind to the complementary polythymidine residues under appropriate conditions (neutral pH, suitable salt concentrations), RNA lacking such poly (A) tails will not. The mRNA and its support can then be washed briefly and the mRNA eluted using a solution of different ionic strength. We will learn why it is important to be able to purify mRNA later.

Now that we have completed our discussion of the techniques for purifying nucleic acids from various sources, test your knowledge of these techniques by attempting the following SAQ.

SAQ 3.1

Match the following reagents (1-9) to the statements (a to g) below. (Use the grid provided).

1) lysozyme; 2) protease; 3) phenol; 4) Na$^+$/ethanol at reduced temperature; 5) RNase; 6) alkaline denaturation and neutralisation with potassium acetate; 7) CsCl density gradient centrifugation; 8) diethylpyrocarbonate; 9) 3 mol l^{-1} sodium acetate.

a) is used to wash nucleic acid pellets to remove DNA leaving the insoluble RNA behind;

b) can be used to precipitate polymeric nucleic acid;

c) can be used to chemically/enzymatically remove protein;

d) can be used to separate chromosomal and plasmid DNAs;

e) acts to stop RNA breakdown by inhibiting the enzymes which destroy RNA;

f) can be used to disrupt bacterial cell walls;

g) can be used to breakdown RNA.

Ring the letter(s) that match each number on the grid below:

1	a	b	c	d	e	f	g
2	a	b	c	d	e	f	g
3	a	b	c	d	e	f	g
4	a	b	c	d	e	f	g
5	a	b	c	d	e	f	g
6	a	b	c	d	e	f	g
7	a	b	c	d	e	f	g
8	a	b	c	d	e	f	g

3.8 Determination of nucleic acid concentration

Before using our purified nucleic acid, we must determine how much we have. The concentration of nucleic acids can be found by ultraviolet absorbance spectrophotometry. The absorbance of light of wavelength 260 nm is directly proportional to the amount of DNA in a solution. The amount of light passing through the solution is compared to that which was incident upon it (see Figure 3.9).

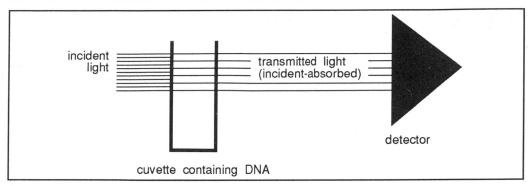

Figure 3.9 Measurement of absorbance.

From this the absorbance can be calculated as follows:

1 absorbance unit = $-\log_{10}$ (transmitted light/incident light). Using light of wavelength 260 nm, 1 absorbance unit corresponds to 50 µg ml^{-1} of double-stranded DNA or 40 µg ml^{-1} of RNA assuming that the light path length of the cuvette is 1 cm.

∏ The use of absorbance measurements at 260 nm is a very convenient method for quickly determining the amount of DNA or RNA present in solution. It is, however, prone to error. Can you think how such errors might arise?

You should have realised that the method assumes that no other material present in the solution absorbs at 260 nm. If, for example, the sample of DNA contains some RNA contamination, then the absorption at 260 nm will not be due solely to the DNA. Therefore using the absorption at 260 nm will over-estimate the amount of DNA present. Similarly proteins also absorb at 260 nm. Therefore, if the DNA or RNA samples are contaminated by proteins, we will over-estimate the amount of nucleic acids present if we rely simply on absorbance at 260 nm. With contaminated samples we have to use more specific assays. DNA can be estimated by reacting the deoxyribose from hydrolysed DNA with diphenylamine. This reaction between deoxyribose and diphenylamine (the Dische reaction) gives a blue product which can be quantitatively determined spectrophotometrically. Similarly, the ribose can be reacted with orcinol which gives a coloured product in the presence of ferric ions. We will not go into details of these alternative methods here. They are described in greater detail in the BIOTOL text, 'Analysis of Amino Acids, Proteins and Nucleic Acids'. For routine work, when we are using purified nucleic acids, the UV absorbance method is usually satisfactory.

SAQ 3.2

You have just isolated some pure plasmid DNA. Before doing your first cloning experiment using this DNA you want to find out its concentration. You make a one hundred fold dilution of your DNA and put 1 ml of the solution into a cuvette of path length 1 cm. The absorbance of the solution is 0.2 absorbance units at 260 nm wavelength. What was the concentration of your original DNA solution?

Ring the correct answer:

1) 1000 µg ml^{-1}

2) 100 µg ml^{-1}

3) 10 µg ml^{-1}

4) 20 µg ml^{-1}

5) 200 µg ml^{-1}

3.9 Checks on purity - the visualisation of nucleic acids and separation on the basis of size

The next stage in *in vitro* genetic manipulation, is to ensure that our isolated nucleic acid is of the desired quality. The sort of questions we would like answers to are:

- is the DNA we have prepared pure or is it contaminated by RNA?

- is the DNA in the correct conformation? For example is it linear or is it in covalently closed circles?

- is the plasmid DNA we have prepared contaminated by linear chromosomal DNA?

- has the chromosomal DNA we have prepared been broken into too many small fragments?

We can couple this check on purity to further purification of the nucleic acids. This is usually done by using a method which separates nucleic acids on the basis of size.

The standard method used to separate, purify and identify DNA fragments is electrophoresis through agarose or polyacrylamide gels. Such techniques are both simple, effective and rapid and the DNA can easily be seen following staining of the gel with ethidium bromide. Both types of gel have a jelly-like consistency and may be made up to different concentrations. The DNA is passed through the gel using an electric field.

3.9.1 Size range of DNAs which are separated

Polyacrylamide gels are effective for separating DNA fragments of size 5-1000 bp (a difference of 1 bp in the size of two fragments is detectable). Table 3.1 gives values for different polyacrylamide concentrations.

Percentage polyacrylamide	Size range of DNA separated (bp)
3.5	100-1000
5.0	80-500
7.5	60-400
12.0	40-200
20.0	10-100

Table 3.1 Concentration of polyacrylamide gel used to separate double-stranded DNA fragments of varying sizes.

Agarose gels in contrast cannot resolve such small size differences but can, however, (depending on the agarose concentration) separate DNAs from 200 bp to 60 000 bp (60 kb) in length (see Table 3.2). Large DNAs up to 10 000 kb long may be separated using a technique called pulsed field gel electrophoresis, in which the direction of the electric field alters periodically. This technique works because large molecules take longer than small ones to change their direction of migration. The run times tend to be of the order of days, but whole chromosomes of yeast can be separated from each other by this technique making it a very powerful tool.

Percentage agarose	Size range of DNA separated (kb)
0.3	5-60
0.6	1-20
0.8	0.8-10
1.0	0.5-7
1.2	0.3-6
1.5	0.2-4
2.0	0.1-2

Table 3.2 Concentration of agarose gel used to separate double-stranded DNA fragments of varying sizes.

3.9.2 Construction of agarose gel

Agarose gels are cast by melting agarose in the desired buffer usually TBE (89 mmol l^{-1} Tris - Borate, 2.5 mmol l^{-1} EDTA) or TAE (40 mmol l^{-1} Tris-base, 5 mmol l^{-1} sodium acetate, 1 mmol l^{-1} EDTA) until a clear solution forms. This can then be poured into a mould in which a horizontal comb can be inserted to form slots (see Figure 3.10) or wells (see also Figure 3.11).

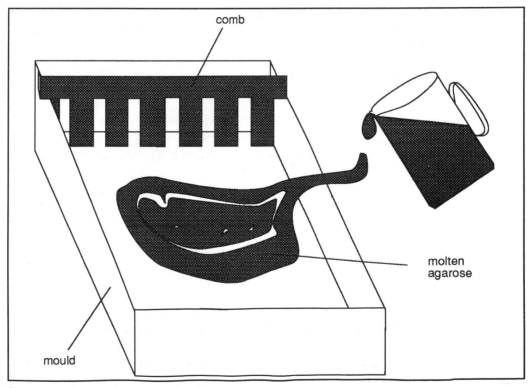

Figure 3.10 Pouring an agarose gel.

When the gel cools, it solidifies and it can then be taken from the casting tray and the comb can be removed. In order to use it to study DNA, it is placed into a tank with reservoirs for buffer at either end, the DNA is mixed with a dye (to track its approximate position during electrophoresis) and a dense solution of glycerol or ficoll so that when it is loaded onto the gel it will sink into the wells of the gel (see Figure 3.11).

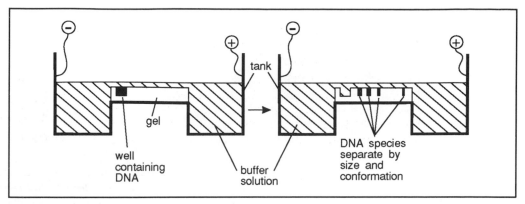

Figure 3.11 Side view of a gel at the beginning and at the end of the electrophoretic run.

The DNA (negatively charged) will migrate towards the anode when an electric field is applied. The distance it moves depends upon its size and conformation OC, CCC or linear (see Figure 3.12).

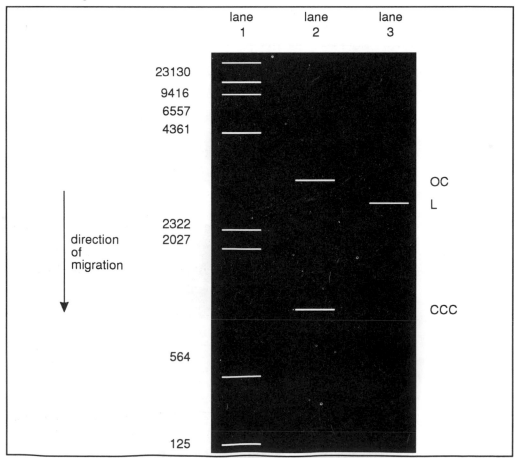

Figure 3.12 Plasmid DNA running on an agarose gel stained with 0.5 μg ml⁻¹ ethidium bromide. Lane 1 : λ *Hind*III restricted size markers (see Section 3.10.1). Lane 2 : open circular (OC) and covalently closed circular (CCC) forms of a 3kb plasmid. Lane 3 : linear 3kb plasmid. Note that the numbers in the left-hand column represent the number of basepairs of linear DNA.

⊓ From the information provided in Figure 3.12 and its legend, does CCC DNA migrate in agarose gels, slower or faster or at the same rate as the corresponding size piece of linear DNA?

Supercoiled (CCC) plasmid DNA is compact in structure and runs through the gel considerably faster than OC DNA plasmid. The relative positions of OC and L plasmid DNA vary depending on whether the gel was run in the absence of ethidium bromide and then stained or run in the presence of ethidium bromide. In the latter case, the concentration of ethidium bromide used effects the migration rates.

Linear DNA fragments are often run on agarose gels in order to size them. The distance migrated from the wells can be plotted against the $\log_{10}$ of the length of DNA (expressed in nucleotide pairs, Figure 3.13).

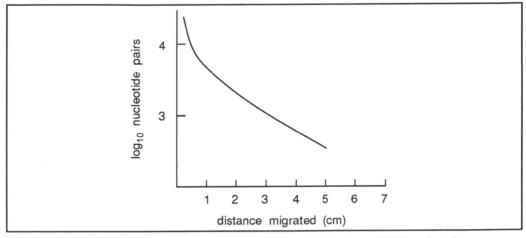

Figure 3.13 An example of a plot of log length (in nucleotide pairs) of a piece of DNA against distance migrated on a 1.2% agarose TBE gel. Note that the actual distance migrated by molecules depends upon the duration of the electrophoresis as well as on the strength of the agarose gel.

TBE has a better buffering capacity than TAE. When TAE is used within the gel electrophoresis apparatus it tends to become exhausted during prolonged electrophoresis (the anode becomes alkaline, the cathode acidic). It should, therefore, be recirculated within the gel elecrophoresis apparatus. For TBE this is not necessary.

⊓ How would you separate single-stranded DNA by electrophoresis?

Obviously you would need to run the electrophoresis under conditions which prevent hydrogen bonds re-forming between DNA strands. Denaturing agarose gels which contain sodium hydroxide may be used to electrophorese single-stranded DNA. Urea may be used in polyacrylmide gels to prevent hydrogen bonds re-forming.

RNA can be run on agarose TBE/TAE gels but better results are often obtained using agarose gels which contain chemicals such as formaldehyde or methyl mercury which denatures the RNA.

3.9.3 Construction of polyacrylamide gels

Polyacrylamide gels are formed following the polymerisation of acrylamide and methylenebisacrylamide and are generally run using the buffer TBE. As the presence of oxygen inhibits the polymerisation process they are cast between two vertical glass plates. Their thickness depends on the thickness of the spacers between the plates. The wells in which the DNA is loaded are formed by a comb inserted at the top (Figure 3.14). The gel is prevented from leaking out from the plates during polymerisation by sealing around the edges of the plates using clamps or tape.

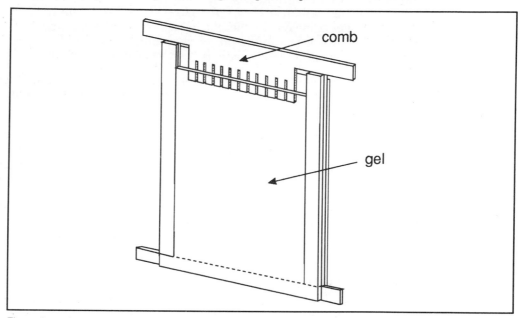

Figure 3.14 Construction of an acrylamide gel.

Small single-stranded DNA fragments can be separated on polyacrylamide gels containing the denaturing agent urea.

SAQ 3.3

What sort of gel would you use if you wanted to:

a) examine a plasmid (size = 5000 bp) you had isolated to see if it was contaminated with chromosomal DNA and RNA?

b) separate two double-stranded DNA fragments sized 15 and 20 bp in length?

c) separate two single-stranded DNA fragments sized 15 and 20 bases in length?

d) separate two chromosomal fragments sized 80 kb and 100 kb?

e) separate 9, 18 and 28s RNA?

The alternatives are:

1) pulsed field gel electrophoresis;

2) non-denaturing 0.8% agarose TBE gel;

3) denaturing methyl mercury agarose gel;

4) 20% non-denaturing polyacrylamide gel;

5) 20% urea containing polyacrylamide gel.

3.10 Genetic manipulation of DNA

Having isolated our pure DNA, the next step is to construct a recombinant molecule, with the desired properties. If, for example, we wanted to transfer a gene for breaking down compound X from *E. coli* to a yeast (for which there is no natural DNA transfer mechanism) we could extract *E. coli* DNA then cut out and isolate the gene of interest and, following many manipulation steps in one or more cloning vectors, transfer it into the yeast via the transformation process that will be described in Chapter 4. This is however, not as simple as it may sound. Such 'cutting' and 'joining' techniques have been developed over the last two decades and involve a number of different enzymes. We will deal with the cutting enzymes first.

vector

We need to explain the term vector. A vector is a 'replicon' into which we can insert genes and other nucleotide sequences. By replicon, we mean a length of DNA that is replicated as a unit from a single initiation site. In genetic engineering, small plasmids, viruses and phage are often used as vectors as they are replicons in their own right. DNA bearing the required genetic characteristics are inserted into these replicons which, when inserted in a suitable host cell are replicated to form multiple copies of the desired DNA sequence. In other words the DNA sequence is cloned. Note that vectors may be constructed artificially. We will discuss these in greater detail in Chapter 4 (see also Section 3.12).

clones

3.10.1. Restriction and modification enzymes

restriction

Restriction/modification systems were first discovered when it was found that some strains of *E. coli* were immune to bacteriophage infection. It turned out that these strains possessed restriction endonuclease enzymes that were capable of cutting foreign DNA at specific sites. Hence phage DNA was degraded when it entered the cell before it had time to replicate and reproduce itself. The bacteria's own DNA is protected against the action of these enzymes due to the presence of additional methyl residues that block this DNA restriction. Such methyl groups are attached to the DNA by modification enzymes. The restriction/modification function is sometimes performed by a single multifunctional enzyme, sometimes by different enzymes. Restriction enzymes can be classified into different types depending on their properties (see Table 3.3).

modification

Property	Type I	Type II	Type III
restriction/ modification	single multifunctional enzyme	endonuclease and methylase separate	separate enzymes with a subunit in common
requirements for restriction	ATP Mg^{2+} S-AM	Mg^{2+}	ATP Mg^{2+} (S-AM)
cleavage site	at least 1 kb from specificity site (possibly random)	at or near specificity site	24-25 bp on 3' side of specificity site

Table 3.3 Characteristics of restriction endonucleases. S-AM = S-adenosyl methionine

Type II restriction enzymes are used in cloning experiments since a particular enzyme will cleave at a known site and nowhere else. The position of the cut is hence entirely predictable. Restriction enzymes are named after the organism from which they were isolated as follows: the first initial of the genus forms the start of the name (capital letter) this is followed by the first two letters of the species (and where applicable a strain designation) finally a roman numeral indicating the order of discovery. For example *Hin* fI was the first restriction enzyme isolated from *Haemophilus influenzae* strain f. The recognition (specificity) sites for type II enzymes are generally tetramers, pentamers or hexamers of basepairs and most have two fold rotational symmetry (are the same when read in either direction on opposite strands, see Figure 3.15).

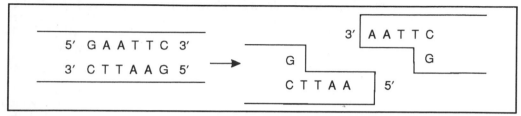

Figure 3.15 Specificity sequence for *Eco*RI. Note that the overhanging ends are produced. In this case they are 5′ overhanging ends.

The convention for drawing these specificity sites is to show just one strand given in the 5′ direction. The position of the cuts can be shown by an arrow.

The *Eco*RI site may thus be written:

5′ G↓A A T T C 3′

∏ Here is a restriction enzyme site:

C T G C A↓G

will this produce 5′ or 3′ overhanging ends?

You should have concluded that the fragments will have 3′ overhanging ends. Thus:

5′ C T G C A G 3′ → 5′ C T G C A 3′   G 3′
3′ G A C G T C 5′   G   3′ A C G T C 5′

∏ What would happen if the restriction enzyme cuts double-stranded DNA in the middle of the restriction enzyme site. For example consider the restriction enzyme site CC↓GG.

You should have concluded that it would produce no over-hanging sequences. Thus

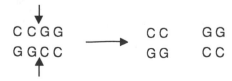

We describe such fragments as having blunt ends.

Some of the common restriction enzyme cut sites are shown in Table 3.4. Many more are given in Appendix 1.

Restriction enzyme	Restriction site
Bam HI	G ↓ G A T C C
Bcl I	T ↓ G A T C A
Bgl II	A ↓ G A T C T
Eco RI	G ↓ A A T T C
Hae III	G G ↓ C C
Hin dIII	A ↓ A G C T T
Mbo I	↓ G A T C
Pst I	C T G C A ↓ G
Sau 3A	↓ G A T C
Sma I	C C C ↓ G G G

Table 3.4 The specificity of some commonly used restriction enzymes. Note that if the restriction enzyme cuts the double-stranded DNA in the centre of the restriction site, then no 5′ or 3′ overhanging ends are produced (compare with Figure 3.15).

isochizomers

So far several hundreds of restriction enzymes have been discovered. Not all enzymes have different specificity sequences, in fact several enzymes recognise and cleave the same nucleotide sequence, such enzymes are called isochizomers.

One unit of enzyme is generally defined as the amount required to completely digest 1 µg of bacteriophage λ DNA in one hour at 37°C in a total volume of 20µl.

Table 2.5 shows some modification enzymes (methylases). Where known, the base which is modified by the corresponding methylase is marked*. We have given more examples in Appendix 2.

Methylase enzyme	Methylation site (*)
Bam HI	G G A T C* C
Eco RI	G A A* T T C
Hae III	G G C* C
Hin dIII	A* A G C T T
Pst I	C T G C A* G

Table 3.5 Specificity of some selected commercially available methylases.

| **SAQ 3.4** | Read the following statements. |

Read the following statements.

1) There are only two classes of restriction enzyme.

2) Only Class 1 restriction enzymes are used in genetic engineering.

3) The cut site of *Eco*RI can be written A $\downarrow$ A G C T T.

4) The methylation site of *Hin* dIII can be written A* A G C T T.

5) If the base marked * in the following sequence A* A G C T T is methylated the enzyme *Hin* dIII will cut it.

6) *Sau*3A and *Mbo* I are isochizomers.

7) Not all enzymes have hexamers as specificity sites.

8) *Pst* I produces fragments bearing 3′ overhanging ends.

9) *Bam* HI, *Bcl* I, *Bgl* II, *Eco* RI, *Hin* dIII, *Mbo* I, and *Sau* 3A give rise to fragments with 5′ overhanging ends.

10) *Sma* I, *Hae* III and *Bam*HI all give rise to fragments with blunt ends.

Now indicate which of these statements are true and which are false.

To help you work these out you may want to draw the double-stranded sequences out and mark on them the cut sites!

In order to cut out gene X (responsible for the breakdown of compound X (see Section 3.10)) from the bulk of the chromosomal DNA, we need to select a suitable restriction enzyme, which will cut either side of our gene of interest and will also cut in a suitable position in the vector we intend to clone it into.

Restriction enzymes will not function unless the correct conditions are provided. The following points should be noted.

- the amount of DNA to be digested must be determined;

- the solution containing the DNA must be adjusted so that it will have the correct ionic composition (NaCl, Mg^{2+} and sometimes K^+) and pH (generally about 7.4) when the enzyme is added;

- incorrect concentrations of NaCl or Mg^{2+} may alter the specificity of the enzyme or decrease its activity. A reducing agent such as dithiothreitol (DTT) is generally included to stabilise the enzyme.

These components are usually adjusted by dilution of a buffer made up at x10 of its normal operational strength. The restriction enzymes can (with some exceptions) be divided into three groups, those that work best in low, medium or high salt conditions and an appropriate 10x buffer used accordingly (see Table 3.6).

Finally the enzyme must be added so that it is at the correct final concentration.

Buffer	NaCl	MgCl$_2$	Dithiothreitol	Tris pH 7.5
low salt	0	10 mmol l^{-1}	1 mmol l^{-1}	10 mmol l^{-1}
medium salt	50 mmol l^{-1}	10 mmol l^{-1}	1 mmol l^{-1}	10 mmol l^{-1}
high salt	100 mmol l^{-1}	10 mmol l^{-1}	1 mmol l^{-1}	10 mmol l^{-1}

Table 3.6 Components of 10x buffer used for selected groups of restriction enzymes.

Figure 3.16 schematically represents the position of a gene of interest (gene X) relative to the restriction sites of several restriction enzymes on a small circular piece of DNA.

∏ Use the information on this figure to develop a strategy for isolating the gene of interest.

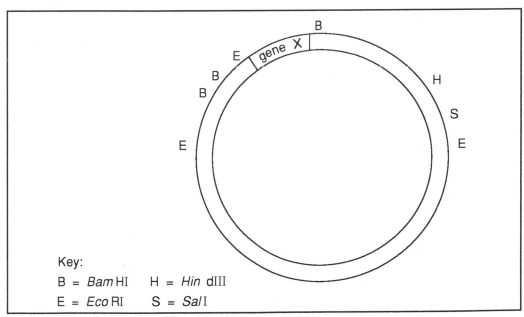

Figure 3.16 Schematic diagram to show the position of gene X with respect to restriction enzyme sites of interest.

If we choose the enzyme *Bam*HI (B on the figure) we could cut the DNA on either side of gene X. If we know the size of the fragments produced by this enzyme we can, after separation of the fragments on an electrophoresis gel, identify the gene of interest by comparison with a set of molecular markers of known molecular weight. The gene of interest may then be extracted from the gel by a variety of means and used in further experiments.

∏ From the information given in Figure 3.16, how many DNA fragments would be produced after treatment with *Bam*HI and on which of these fragments would we find gene X.

You should have concluded that as the DNA would be cut at each B site, three fragments would be produced. One would be small, another a little larger containing an E site and gene X and a large fragment carrying H, S and two E sites. Gene X would be therefore on the intermediary sized fragment.

This example is a simplification of real situations since we first have to determine the position of the gene of interest relative to the restriction enzyme sites. We will enlarge on this later in this text. It is, however, sufficient at this stage to grasp the overall strategy.

SAQ 3.5

Below is a map of a piece of DNA of interest (in this case it is a plasmid), the numbers refer to the nucleotide position starting from the *Eco*RI site (0/3000). All nucleotides are numbered from the nucleotide at the centre of the *Eco*RI(E) cut site.

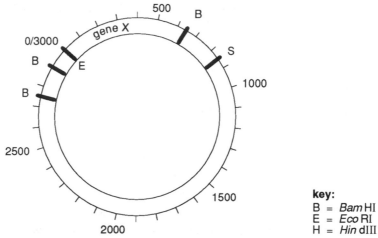

key:
B = *Bam* HI
E = *Eco* RI
H = *Hin* dIII

A gel is run which contains:

lane 1 λ *Hin* dIII markers (contain fragments of DNA of known size); lane 2 uncut plasmid DNA; lane 3 *Sal*I (S) cut plasmid DNA; lane 4 *Bam*HI (B) cut plasmid DNA.

Which is the smallest fragment containing gene X?

	lane 1	lane 2	lane 3	lane 4
23130	—			
9416	—			
6557	—			
4361	—			
		— a		
			— c	
2322	—			
2027	—			— d
		— b		
564	—			— e
125	—			— f

3.11 Manipulation of the fragment containing DNA of interest

Having isolated our fragment of interest, what do we take as our next course of action? We can use it as it is, further restrict it, or manipulate it using other enzymes.

Our decision depends upon many factors:

- do we need to remove excess DNA for any reason (perhaps to remove an additional gene we do not want)?
- do we need to add a new promoter (to ensure the gene will function in its new host)?
- is it of suitable size to fit into our chosen cloning vector?
- are its restriction sites suitable?

In order to deal with these issues we will now consider the other enzymes that we can use to manipulate the DNA.

3.11.1 Enzymes used to manipulate DNA fragments

Some of the enzymes commonly used in cloning are shown below:

Nucleases (enzymes which cut/degrade DNA)
restriction endonucleases (with specific recognition sites)
endonucleases (without specific recognition sites)
exonucleases
Ligases (DNA joining enzymes)
Phosphatases and Kinases (enzymes which remove or add respectively phosphate groups)
DNA synthesising enzymes

Table 3.7 Enzymes commonly used in molecular biology.

To understand how the enzymes summarised in Table 3.7 work we have to be aware of the structure of DNA. Up until now we have considered the molecule as a series of nucleotides represented for example in the case of the *Eco*RI specificity site as:

$$5' \quad G \ A \ A \ T \ T \ C \quad 3'$$

$$3' \quad C \ T \ T \ A \ A \ G \quad 5'$$

But what is not illustrated quite so clearly in such a diagram is the nature and polarity of the molecule. Each nucleotide is linked to the previous, such that the 5' end of each molecule has a phosphate group (P) and the 3' end bears a hydroxyl group (OH).

This can be shown in a simple form as:

$$5' \quad _p G _p A _p A _p T _p T _p C _{OH} \quad 3'$$

$$3' \quad _{OH} C _p T _p T _p A _p A _p G _p \quad 5'$$

with this information in mind we will now look at the mechanisms of action of the different enzymes used.

In the following sections, we will describe the main classes of enzymes used in genetic engineering. For reference purposes, we have provided lists of specific examples in Appendices 1 and 2. You will find these useful to refer to if you are seeking an enzyme with a particular property.

3.11.2 Nucleases

We have already discussed the importance of restriction enzymes and will give a further example of their use in analysing cloning products later. Restriction endonucleases are not the only enzymes which cut DNA. Other endo-exonucleases, which lack specific recognition sites can also be used to alter DNA.

Exonucleases

Exonucleases remove DNA nucleotides by 'nibbling' away at the ends of pieces of DNA. They can either act on both strands or just one. There are many examples of endo- or exonucleases, some act on DNA with a 5' protruding terminal phosphate, some on 3'-OH ends, some on double-stranded and some on single-stranded DNA.

Endonucleases

Endonucleases cut within a DNA molecule, they can be specific for either double-stranded or single-stranded templates. The positions in which they cut DNA is, unlike the restriction enzymes, not specified by the sequence of nucleotides.

Multifunctional nucleases

Some nucleases are multifunctional and exhibit both endo- and exonuclease activities. For example the enzyme *Bal* 31 exhibits two activities:

- an endo- and exonuclease activity that brings about the removal of oligonucleotides (short chains of nucleotides) or mononucleotides from the 5' and 3' termini of double-stranded DNA;
- a single strand specific endonuclease.

Having cut and trimmed our piece of DNA, we can now consider inserting it into our vector molecule using the joining enzyme ligase.

3.11.3 Ligase

T4 DNA ligase

The enzyme ligase which is used in molecular biology is usually purified from *E. coli* which have been infected with the phage T4. *In vitro* the enzyme acts to repair discontinuities in DNA strands. This is illustrated in Figure 3.17.

It will also join compatible ends of the same molecule and is used to clone pieces of DNA into plasmid vector molecules (Figure 3.18).

In Figure 3.18 we show our gene of interest to be cloned into a vector. This requires the ends of the molecules to be compatible. For example in SAQ 3.5 we cut out our gene X from our donor plasmid using enzyme *Bam*HI. If you look at Figure 3.17 (ligation of cohesive ends) closely you will see that we illustrate this using cohesive ends derived from *Bam*HI restricted DNAs. Had we chosen *Bam*HI to cut out our gene X and an enzyme such as *Pst*I to cut our vector (see Table 3.4) we would have been unable to ligate the two together since the bases in the overhanging strands would have been incompatible.

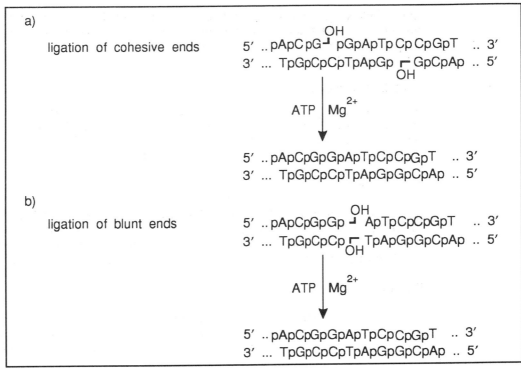

Figure 3.17 Action of T₄ DNA ligase. A phosphodiester bond is formed between adjacent 5′ phosphate and 3′OH groups. This can be achieved using DNA with over-hanging (cohesive) ends a) or with blunt ends b).

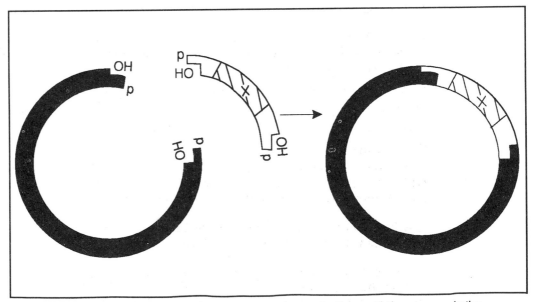

Figure 3.18 DNA can be ligated into a plasmid which has been cut with a restriction enzyme. In the example shown gene X is inserted into a plasmid.

3.11.4 Adaptors, linkers and homopolymer tailing reactions

As ligases can only join molecules together but cannot fill in gaps where nucleotides are missing not all cloning operations are straightforward. We can get around these problems however by the use of adaptors, linkers and 'tailing' reactions and by using DNA synthesising enzymes (Section 3.11.6). We will deal with adaptors, linkers and tailing reactions in this section.

Adaptors, linkers and tailing reactions are means of introducing compatible 'sticky' termini onto molecules.

linkers   A linker is a small sequence of nucleotides containing a known restriction enzyme site(s) that can be used to link two pieces of DNA together. Figure 3.19 shows how we may use linkers to introduce such termini. As we can see, many linkers may ligate to the blunt ended piece of DNA we wish to clone. The excess can be removed using the restriction enzyme whose specificity site is contained within the linker (in this case *Eco* RI). The blunt ended DNA molecule we are cloning must (if it has any internal *Eco* RI sites) of course be treated with *Eco* RI methylase to prevent it being cleaved by the restriction enzyme. The molecule we have produced at the bottom of Figure 3.19 can be linked to other pieces of DNA with *Eco* RI compatible (adhesive) ends using T4 DNA ligase.

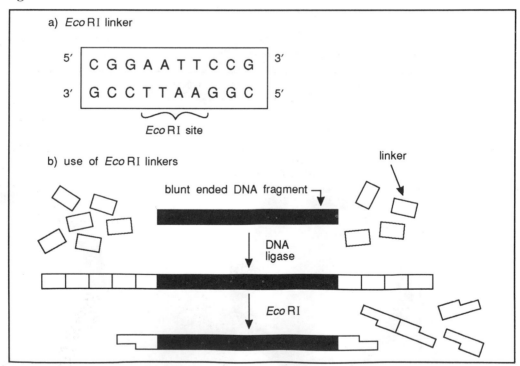

Figure 3.19 Linkers and their usage.

Adaptors are short sequences of nucleotides similar to linkers except that they may already have 'built in' adhesive ends and hence can avoid the necessity for such additional enzyme treatments. If you study Figure 3.20, you can see that the adaptor can be synthesised so that the overhanging cohesive 5′ terminus lacks a phosphate group and hence is incapable of ligation into long chains (compare Figures 3.19 and 3.20) (we will examine this in more detail in Section 3.11.5).

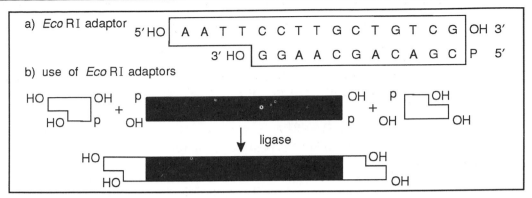

Figure 3.20 Adaptors and their usage.

Homopolymer tailing

homopolymer tails

A general method for joining DNA molecules is the use of homopolymer tailing (see Figure 3.21). In this technique, the enzyme terminal deoxynucleotidyl-transferase is used to add tails of a single deoxynucleotide type to the overhanging 3′ terminus of a molecule such as that produced by the enzyme *Pst*I. By tailing one partner with As and the other with Ts (or Cs and Gs respectively) the molecules can be 'annealed', any missing nucleotides can be filled in using DNA polymerase and the molecule covalently closed by ligase (the term 'anneal' means that we allow complementary bases to pair and form hydrogen bonds).

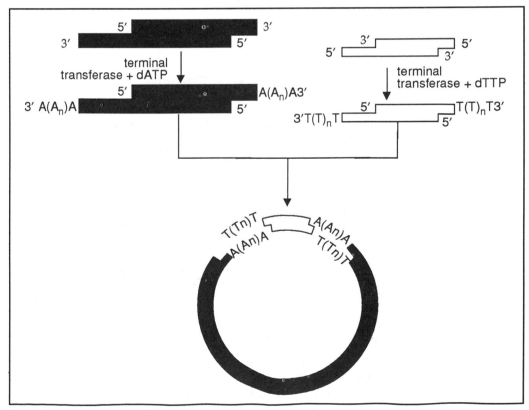

Figure 3.21 Homopolymer tailing (see text for details).

3.11.5 Phosphatases and kinases

As we explained in Figure 3.17, ligase works by putting a phosphodiester bond between an adjacent 5' terminal phosphate and an adjacent 3' OH group. By synthesising adaptors which lack this phosphate group on their 5' overhanging termini we can minimise unwanted ligation products. Unwanted 5' terminal phosphate groups can be moved enzymatically.

Alkaline phosphatase

The enzyme alkaline phosphatase acts to remove the 5' terminal phosphate from a DNA molecule leaving an OH group. Figure 3.22 shows how treatment of vector DNA with alkaline phosphatase, prior to ligation of a gene into it, can prevent recircularisation of vector DNA and formation of vector dimers and trimers.

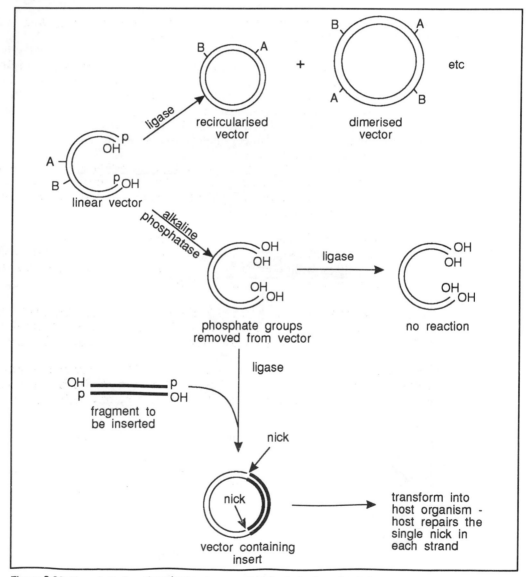

Figure 3.22 Use of alkaline phosphatase to prevent recircularisation of a vector.

The top left-hand section of Figure 3.22 shows a circular vector cut by a restriction endonuclease to produce a linear vector carrying 5' phosphate groups. If this is incubated with ligase, it can re-circularise or two or more vector molecules can be joined together to form dimers or trimers. Incubation with alkaline phosphatase removes the 5' phosphate groups and such molecules cannot recircularise when incubated with ligase. On the other hand, if such vectors are incubated with fragments carrying 5' phosphate groups, ligase will attach these fragments to the vector by forming phosphodiester bonds between the 5' ends of the fragment and the 3'OH ends of the vector. Notice that there is no linkage between the 5'OH ends of the vector and the 3'OH ends of the DNA fragment. Thus a 'nick' (lack of joining) occurs in these positions. These can, however, be repaired if this recombinant DNA is introduced into a host cell.

Polynucleotide kinase

polynucleotide kinase

The enzyme polynucleotide kinase will add a phosphate group to a free 5'-terminus hence reversing the effect of alkaline phosphatase.

As we have explained we can add or remove the phosphate groups which are crucial to the function of the enzyme ligase. Ligase however can only join adjacent nucleotides, it is unable to synthesise DNA to fill in any gaps that are present. The role of DNA synthesis is fulfilled by enzymes called polymerases.

DNA polymerases

DNA polymerase I

DNA polymerase I attaches to a short single-stranded region in a mainly double-stranded DNA molecule. Here it fills in the gap and replaces the existing nucleotides (see Figure 3.23a)). The enzyme exhibits dual activity; it both acts as a $5' \rightarrow 3'$ polymerase and as an exonuclease.

Klenow fragment

The enzyme can be cleaved to yield a large and a small fragment. The large fragment is called the Klenow fragment and is notable in that it retains the polymerase function but lacks the $5' \rightarrow 3'$ exonuclease activity. Hence the Klenow fragment will fill in a gap but leave existing nucleotides (see Figure 3.23 b)).

The Klenow fragment is often used to fill 5' overhanging cohesive termini resulting from restriction digests prior to the addition of linkers or adaptors, or to convert incompatible 5' overhanging ends to blunt ends prior to ligation (Figure 3.23 c)).

reverse transcriptase

Reverse transcriptase is unique in that it uses an RNA template to synthesise a cDNA (complementary DNA) strand which is complementary to the RNA (see Figure 3.23 d)). This is important for a technique called cDNA cloning which is discussed in a later chapter.

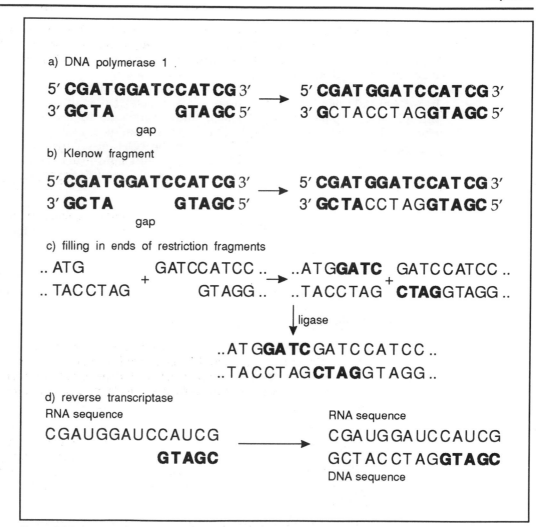

Figure 3.23 The activities of DNA synthetic enzymes.

As Sections 3.10 and 3.11 were quite complex it is recommended that you should read through them again before attempting the following SAQ.

SAQ 3.6

You wish to clone gene X which is located on a *Bam*HI fragment into the vector drawn below. The recombinant molecule will then be introduced into a strain of *E. coli* which is sensitive to ampicillin (amp S) and tetracycline (tet S) and grown on agar plates.

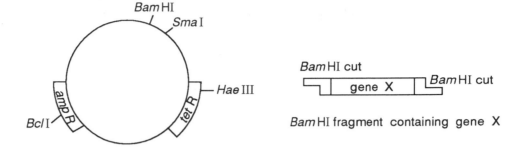

Note *amp* R = ampicillin resistance gene
          tet R = tetracycline resistance gene

Which of the following would be possible?

1) Cut the vector with *Bam*HI and ligate the *Bam*HI fragment containing gene X into it. Bacteria which contain the plasmid with gene X inserted into it may be identified by their ampicillin/tetracycline-resistant nature. Restriction of the plasmid DNA gives two fragments, the plasmid and the *Bam*HI fragment containing gene X.

2) Clone the *Bam*HI fragment containing gene X into the *Bcl* I site of the vector (as the restriction fragment ends are compatible). Select bacteria containing plasmids with an inserted gene X by their resistance to tetracycline, but sensitivity to ampicillin. Insert containing plasmids upon restriction with *Bcl* I will give two fragments, the plasmid and the *Bcl* I fragment containing gene X.

3) Convert the ends of the fragment to blunt ends using a DNA synthesising enzyme or an exonuclease, clone into the *Hae* III site of the vector. Bacterial colonies containing plasmids with an insert should be tetracycline sensitive but ampicillin resistant.

4) Convert the ends of the fragment to blunt ends using a DNA synthesising enzyme or an exonuclease, add *Bcl* I adaptors and clone into the *Bcl* I site of the vector. Bacterial colonies containing plasmids with an insert should be tetracycline resistant but ampicillin sensitive. Restriction of plasmid DNA with the enzyme *Bcl* I should give rise to two fragments, the plasmid and the filled-in *Bam* HI fragment containing the gene with *Bcl* I adaptors on its ends.

Select the strategy which would be most successful in introducing gene X into this *E. coli* strain.

Having successfully learnt how to isolate nucleic acids and how to clone fragments of interest, we now have to be able to decide which vectors will be the best for our purposes. then we have to reintroduce our vector into a host in which it can replicate itself and be able to identify those host organisms containing our recombinant vector molecules.

3.12 Cloning vectors

As cloning vectors and the means of reintroducing them into a suitable host are dealt with in Chapter 4 we will simply summarise the types of cloning vehicle available here.

3.12.1 The ideal cloning vector

The ideal cloning vector:

- is of low molecular weight, so that it is easy to handle and isolate;

- will confer a readily selectable phenotype on host cells carrying it, such as antibiotic resistance, or contain a gene whose product is easily recognisable given the correct conditions;

- will have a single site for a large number of restriction endonucleases, preferably in a gene with a readily identifiable phenotype.

∏ It would be worthwhile to see if you can work out the reasons why a single site for a large number of restriction endonucleases in a gene which produces readily identifiable phenotype is desirable in a cloning vector.

You should have reasoned as follows:

Let us assume that we have a gene (gene X) which bestows an early identifiable phenotype (for example it could be an antibiotic resistance or an ability to use a particular substrate).

If we insert a new gene within gene X then gene X will become non-functional.

Thus:

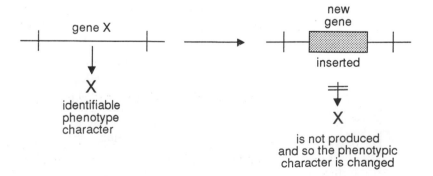

This enables us to identify cells which are carrying genes into which new nucleotide sequences have been inserted. If gene X contains single sites for a large number of restriction endonucleases, then it means we can make single cuts in gene X with different 'overlapping' ends. This in turn means we can use a wide variety of restriction enzymes to isolate DNA fragments and that their cohesive ends will be complementary to the cut ends of gene X. Therefore, we can conveniently use such a vector for cloning a very large number of different nucleotide sequences. Diagrammatically we can illustrate this by:

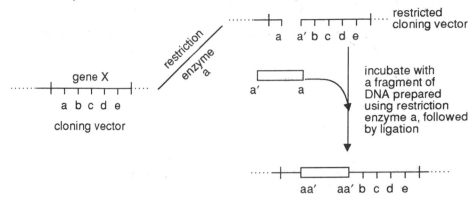

Similar procedures could be followed using restriction enzymes b,c,d and e and pieces of DNA carrying the equivalent overlapping (cohesive) ends.

3.12.2 Cloning vectors based on *E. coli* plasmids

There are a large number of commercially available, small plasmid vectors with easily selectable markers. An example of such a commonly used vector is pUC18. This is a high copy number plasmid which contains an ampicillin resistance gene and a means to select for the presence of a cloned gene (see Section 3.13).

3.12.3 Plasmid cloning vectors for use in other bacteria

Most *E. coli* vectors are unsuitable for use in commercially important organisms such as *Streptomyces* and *Bacillus* species as they are unable to replicate themselves in these hosts. Special vectors with similar requirements for use (suitable means of selection and screening, specific unique restriction sites) have been constructed which can replicate themselves in a broad range of bacterial hosts.

3.12.4 Vectors capable of carrying large DNA fragments

Plasmid vectors are unable to carry the large fragments of DNA which are sometimes needed in cloning experiments. On such occasions, phage, cosmid and Y.A.C. vectors can be used. These are dealt with in the next chapter.

3.13 Selecting plasmids which contain a cloned DNA fragment

Having decided which vector is the most suitable for our purposes, it is important to be able to determine whether or not we have been successful in cloning our gene within the vector.

When a mixture of DNA fragments containing 5′ phosphate groups are incubated with ligase, it is possible to join the fragments in a variety of combinations. For example, consider the mixture of a linearised vector and a desired gene X. Assuming that the two types contain compatible cohesive ends (for example they were prepared using the same restriction enzyme) then it is possible to produce such combinations as:

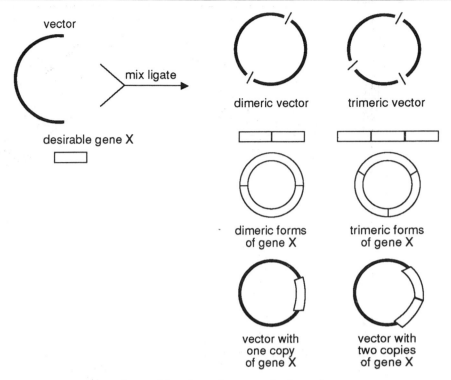

There are many more such combinations that may be found.

It is important to be able to have a mechanism that enables us to identify plasmids which contain our cloned fragment of interest. A number of commercial vectors are available which possess such characteristics. We will use pUC18 as an example. It is illustrated in Figure 3.24. pUC18 has the following characteristics:

- it contains a synthetic series of restriction enzyme recognition sites, closely arranged in what is known as a polycloning or multi-cloning site. These restriction sites are unique and not found anywhere else in the plasmid. There are 13 cleavage sites in total in this short section, these include sites for *Hin* dIII, *Pst* I, *Bam*HI, *Sma* I and *Eco* RI;

- this polycloning site is incorporated into a gene which encodes a biologically active fragment of the enzyme β galactosidase. Insertion of a fragment of DNA into one of the cloning sites within this gene can inactivate it. Such recombinant clones lose the ability to hydrolyse a colourless chromogenic substrate into a blue readily visible product. Colonies without such an insert are therefore blue, those with the inactivated gene are white, in the presence of the substrate.

Other methods of selecting for the presence of a cloned fragment can be used, for example the inactivation of an antibiotic resistance gene such as those coding for ampicillin resistance or tetracycline resistance.

Complex vectors which allow the production of single-stranded DNA and RNA products may also be used. So may vectors which permit the high level expression of foreign proteins. Discussion of these vectors is however beyond the scope of this chapter and will be dealt with in later sections. Also the vectors available for specific groups of organisms are described in detail in the BIOTOL text, 'Strategies for Engineering Organisms'.

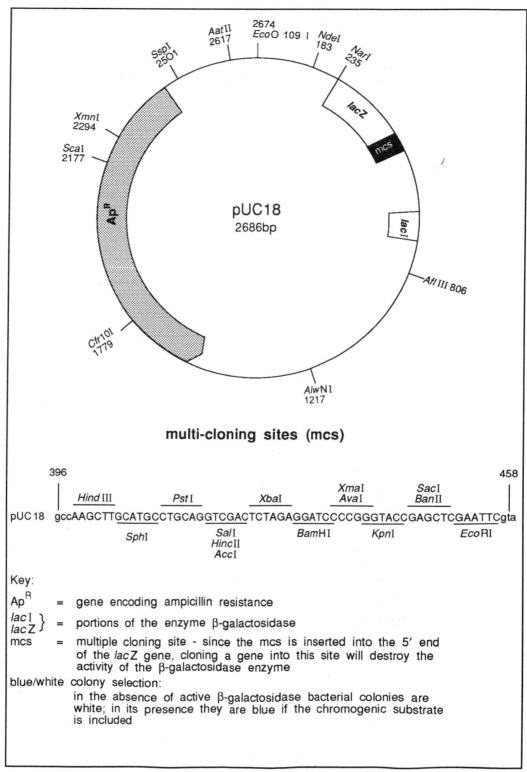

multi-cloning sites (mcs)

Key:

Ap^R = gene encoding ampicillin resistance

lac I
lac Z } = portions of the enzyme β-galactosidase

mcs = multiple cloning site - since the mcs is inserted into the 5′ end
of the *lac* Z gene, cloning a gene into this site will destroy the
activity of the β-galactosidase enzyme

blue/white colony selection:
in the absence of active β-galactosidase bacterial colonies are
white; in its presence they are blue if the chromogenic substrate
is included

Figure 3.24 pUC18. The restriction enzymes which cut once only are shown on this plasmid map.

3.14 Screening by hybridisation

In the screening of constructs by hybridisation we make use of the fact that DNA and RNA can be transferred to specially prepared nitrocellulose or nylon membranes and immobilised there. Such fixed nucleic acids can be detected by hybridisation. This technique is frequently used to identify DNA/RNA sequences of interest.

3.14.1 Immobilisation of phage or bacterial DNA on a membrane.

Immobilisation of DNA on a cellulose or nylon membrane is quite straightforward. In the example illustrated in Figure 3.25, for the immobilisation of phage DNA, the membrane is simply pressed onto phage plaques in the 'lawn' of bacteria growing on a nutrient agar surface. Some of the phages are transferred to the membrane. By exposing the membrane to suitable conditions (for example by raising the pH to 12-12.5) the DNA is denatured. After neutralising the single-stranded DNA is 'fixed' to the membrane. This is usually achieved by heating it.

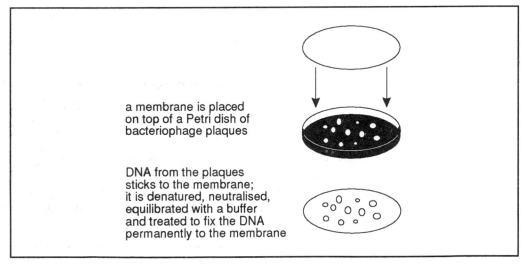

a membrane is placed
on top of a Petri dish of
bacteriophage plaques

DNA from the plaques
sticks to the membrane;
it is denatured, neutralised,
equilibrated with a buffer
and treated to fix the DNA
permanently to the membrane

Figure 3.25 Immobilisation of phage DNA on a membrane.

immobilisation of bacterial DNA

The same technique may be used for bacterial colonies except that after contact of the membrane with the colonies it is placed inverted (bacterial side uppermost) onto a nutrient agar plate and incubated for a few hours to allow growth of the bacteria into fresh colonies (identical in spatial orientation to the original plate). The filter is then treated to lyse the bacteria and release the DNA, this is followed by denaturation, neutralisation and equilibration in buffer.

3.14.2 Immobilisation of DNA/RNA from agarose gels

Southern and Northern blotting

In the same way that it is important to be able to transfer DNA from plaques and bacterial colonies onto a solid support, it is also necessary to be able to immobilise DNA or RNA from an agarose gel. The process of transferring DNA is known as Southern blotting and that of RNA is known as Northern blotting.

The methodology is very simple and first involves the denaturation of the nucleic acid. In the case of DNA, we can treat the gel with NaOH followed by neutralisation. There is no need for such treatment of an RNA gel as the gel itself includes the denaturing agent (see Section 3.9.2) when it is run.

In order to transfer the nucleic acid, the gel is laid on top of a pad of buffer-soaked filter paper. Then its top surface is covered with a similar sized piece of nitrocellulose or nylon membrane. Overlaid on the top is a pile of filter paper (see Figure 3.26). The buffer, which is drawn up through the gel (by the dry filter paper) carries with it the RNA or DNA fragments. When the nucleic acid contacts the membrane, it binds and can be fixed by baking at 80°C (or if preferred by UV treatment in the case of nylon).

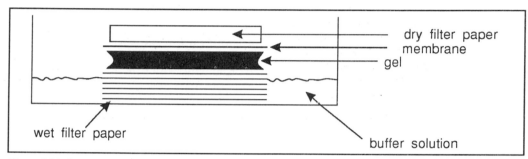

Figure 3.26 Southern/Northern blotting.

3.14.3 Hybridisation as a means of detecting sequences of interest.

autoradio-
graphy

Hybridisation is a technique in which a short RNA or DNA probe, whose nucleotide sequence is complementary to a sequence of interest, is allowed to anneal (hybridise) under the appropriate conditions to denatured, single-stranded DNA or RNA. The probe will hybridise with its complementary sequence, that which has not bound is washed off and the position of the bound probe is detected in some manner. This is often achieved by incorporating radioactive nucleotides into the probe prior to hybridisation. The presence of these can then be determined by exposing a piece of autoradiography film next to the hybridised membrane. The film will blacken when radiation from the probe hits it. This is illustrated in Figure 3.27.

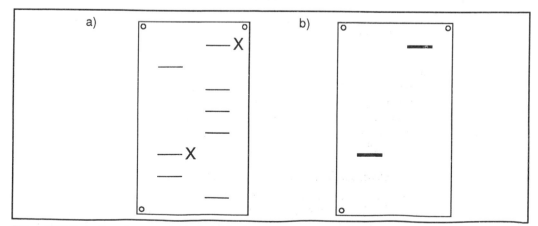

Figure 3.27 Detection of sequences by hybridisation. a) Shows an agarose gel containing a number of DNA fragments on it. X marks the bands containing our gene X of interest. b) Shows an exposed autoradiography film. The radiation from the probe, complementary in sequence to gene X, has blackened the film and we can detect the position of the band containing gene X.

In the same way a colony or plaque containing a sequence of interest may be detected and so distinguished from the colonies or plaques not containing the sequence of interest.

3.15 Concluding remarks

We have now covered many of the basic *in vitro* techniques required for the isolation, cloning and modification of genes and identification of sequences of interest.

There are of course many questions we have not, as yet answered. For example, how do we prepare RNA and DNA probes? Will the genes we have inserted into vectors be expressed? Nevertheless you should appreciate what you have already learnt. The techniques we have described and refinements arising from them, enable us to transfer genes between quite unrelated organisms. They also enable us to produce hybrid genes, by which we mean that we can construct new genes from parts of existing genes by joining them together. If such genes are expressed, we would expect them to code for recombinant proteins. Note that we have particularly focused on the use of cloning vectors in this chapter. Cloning vectors are vectors that enable us to make multiple copies of nucleotide sequences (genes). There are however other types of vectors called expression vectors. These are designed so that our newly inserted DNA sequences are expressed. We will be examining some of these later. The principles underpinning their design and manipulation have much in common with the cloning vectors.

Many biotechnologically important genes have been manipulated using these methods and have been successfully used to produce recombinant proteins. Such genetic manipulations have had many diverse applications which range from the production of human insulin, which is approved for clinical use in the treatment of diabetes, to changing petunia flower colour following the introduction of a gene from maize.

In the chapters that follow we will enlarge on the basic understanding you have gained from this chapter.

Summary and objectives

In this chapter we have provided you with a description of the core techniques used in *in vitro* gene manipulation (DNA technology). We have considered how suitable sources of genetic material may be extracted, purified and characterised and explained how material may be inserted and cloned in suitable vectors.

Now that you have completed this chapter you should be able to:

- explain the principles involved in the extraction of total cellular DNA from bacteria, plant, animal and yeast cells and appreciate the common aspects of such procedures;

- explain how to extract bacterial plasmid DNA and describe two methods of how to separate it from other cellular components;

- list the steps involved in bacteriophage DNA isolation;

- describe a method to isolate and purify RNA and list the precautions that should be undertaken to avoid RNases breaking down the RNA during isolation;

- use optical density measured at 260 nm to calculate DNA concentrations;

- compare and contrast the methods to separate and visualise nucleic acid on a basis of size and conformation;

- select the method of choice to separate DNA/RNA of different sizes;

- recognise the specificity sequences and be able to draw the termini following restriction, of the following restriction endonucleases; *Bam* HI, *Bcl* I, *Eco* RI, *Hae* III, *Hin* dIII, *Mbo* I, *Pst* I, *Sau* 3A and *Sma* I;

- appreciate how methylases modify DNA and their effect on restriction endonuclease activities;

- realise that restriction endonucleases have specific requirements to retain their activity;

- determine the approximate sizes of double-stranded DNA fragments run on an agarose gel alongside known size markers;

- describe the function performed by: endo/exo nucleases, ligase, phosphatase, kinase and DNA synthesising enzymes of different types;

- given appropriate information, determine a strategy to clone a DNA fragment into a simple vector;

- list the important features of a cloning vector;

- understand the principles of gene inactivation in determining whether or not a vector contains an insert;

- describe how to transfer DNA from a bacterial colony, a phage plaque or an agarose gel onto a nylon membrane;

- explain the principles involved in the identification of a sequence of interest by hybridisation.

Introduction of DNA into living systems

Introduction of DNA into living systems

4.1 Introduction

If we wish to introduce DNA into a new host cell, we are faced with two key problems:

- the DNA must be able to physically cross the cell membrane and, in the case of walled organisms, some means of passing or avoiding this extra barrier must be found;

- once inside the new host cell, our DNA must be able to replicate and, in the case of most eukaryotes, integrate into the host genome. Additionally, we should be able to apply some type of selection pressure to maintain our DNA throughout cell growth and division.

This chapter discusses the way these two problems may be overcome.

In Chapter 2 we learnt that exogenous DNA can be taken up by bacterial cells by the process of transformation. We also learnt that DNA can be packaged into bacteriophages and that this DNA can be transmitted to new host cells by the process of transduction. These are therefore two natural methods for gene transfer ready for us to exploit. In this chapter we will explain how this is achieved. The key to the success of introducing new DNA into living systems is to ensure that the DNA is in such a form that it will be successfully replicated in its new host cell. Thus, much of this chapter is concerned not simply with getting the DNA into the cell but with ensuring that it is in a suitable form for replication.

4.2 Transformation

4.2.1 Getting DNA into the cell

We will now look in a little more detail at the first of our problems. How can we get DNA into the cell? There is an enormous range of techniques available but only a limited number of methods will be suitable for any one organism. Here we will cite the main methods. Later, we shall consider more detailed examples using bacteria and simple eukaryotes to illustrate general principles. The main methods are listed below:

- **The use of cells which are naturally competent** for uptake of naked DNA. Amongst bacteria *Bacillus subtilis* is notable for this feature and the optimum transformation conditions are well understood.

- **Calcium treatment of cells** of *E. coli* allows transformation of plasmid DNA at high frequencies. Shock treatments with a range of other metal ions also improves the efficiency even further.

- **Transformation of protoplasts** of walled organisms like fungi, higher plants and some bacteria. With walled organisms, the main barrier to introducing DNA into the cells are the cell walls themselves. Such cell walls can be removed by hydrolysing them using suitable enzymes. For example, with bacteria, the enzyme lysozyme will hydrolyse the peptidoglycan component of the cell wall. Similarly the enzymes cellulase and pectinase will hydrolyse cell wall components of plant cell walls and chitinase will hydrolyse the chitin components found in some fungal cell walls. Alternatively, with larger cells (mainly plant cells), mechanical methods can be used to prepare protoplasts (details are provided in the BIOTOL texts, '*In Vitro* Cultivation of Plant Cells' and 'Strategies for Engineering Organisms'). Removal of the cell wall produces fragile protoplasts. They must be maintained in isotonic solutions otherwise they take up water, swell and burst. The protoplasts prepared in this way are suitable for introducing exogenous DNA into the cytoplasm. A variety of techniques have been used to facilitate the uptake of DNA. This includes the use of polyethylene glycol (PEG) and electroporation (see below). There are two fundamental problems with using protoplasts. These are, the fragility of the protoplasts and problems encountered in the regeneration of whole cells after the exogenous DNA has been taken up. You should also note that polyethylene glycol is toxic at high concentrations although different cells display different sensitivities to this chemical. Thus polyethylene glycol concentrations need to be carefully evaluated in terms of transformation efficiency and host cell survival.

enzymatic removal of cell walls

fragility of protoplasts and transformation efficiency

- **Electroporation of cells/protoplasts** using high voltage shocks for a fraction of a second to make membranes more permeable to DNA. Holes are fleetingly produced in the plasma membrane, allowing DNA entry. A proportion of such cells will be stably transformed even in animal cell lines where no other technique will work.

- **Microprojectile bombardment** of intact cells with metal particles (eg tungsten; diameter ≤ 4 μm) coated with DNA. This method does not require complex cell cultures or pretreatments and avoids the problems of regenerating protoplasts.

- **Infection of cells with viruses** resulting in transfer of genes from the virus to the host cell.

- **Microinjection** of DNA directly into a cell nucleus. This is a standard method for introducing genes into animal cell lines. Newly fertilised eggs are transformed in this way *in vitro*, prior to being transplanted into the mother's womb at an early stage of embryo development.

4.2.2 Selection of transformation technique

You will realise from this list that we have potentially a wide variety of techniques available to us. It should be apparent from the descriptions of these techniques that they are not all equally applicable. Some work well with whole cells, others require considerable pre-preparation (for example in production of protoplasts) before the DNA can be introduced. Some methods require specialist equipment (for example in electroporation, microprojectile bombardment and micro-injection). With some of the methods, we can use a large number of cells (for example when using naturally competent cells) whilst in others the technique is labour intensive and can only be carried out with a limited number of cells (for example micro-injection).

not all methods are equally applicable

Ultimately we would like to be able to select the most appropriate method. It would help you to be able to make the most suitable choice by thinking about the advantages and disadvantages of each of the techniques listed. The following exercise will help you to achieve this comparison.

Π Draw up a table listing the features of the techniques described above using the following headings:

Technique	How much pre-preparation of the cells is required?	Is specialist equipment needed?	Can the method be used to tansform a lot of cells?	After DNA transfer, are there any difficulties in recovering viable cells?

This should also help you realise that the efficiency of each of these techniques in terms of the number of cells (protoplasts) transformed also depends upon the cell types being used. The choice of technique, therefore, depends upon previous experience, the availability of equipment and upon informed guesswork.

Usually, when transformation is first attempted for a 'new' organism, the efficiency of transformation is determined for a variety of methods using a range of variants of these basic techniques (for example, we might use a range of salt concentrations or change the ratio of DNA:cells).

transformation efficiency

Transformation efficiency can be expressed in a variety of ways such as the number of transformants per μg of added DNA or the number of transformants per viable cell. One common problem when we try to transform protoplasts is that many protoplasts will not regenerate to the normal cell form. In these cases, in order to get a realistic idea of the true transformation efficiency per viable cell, we have to score the protoplast regeneration frequency.

In any transformation procedure, we find that many, or most, of the viable cells will not have taken up DNA.

Π How then can we detect those cells which are true transformants?

importance of selectable markers

The easiest way is to ensure that our transforming DNA carries a selectable marker gene which confers some easily scored character upon transformants. Such markers could be, for example, a gene coding for antibiotic resistance.

SAQ 4.1	The plasmid pIJ702 is used to transform protoplasts of the filamentous, antibiotic-producing bacterium, *Streptomyces*. Using a standard PEG-mediated system, you attempt to transform two strains (of two different species); *S. coelicolor* and, *S. hygroscopicus*. Using 2μg of DNA, in each case, you obtain 10^7 transformants from *S. coelicolor* and 25 from *S. hygroscopicus*. If you used 10^9 protoplasts in each case with a viability of 25% what is the efficiency of transformation:

1) in transformants per μg added DNA;

2) in transformants per viable protoplast?

4.2.3 An example of a protocol for transformation

Below we provide the details of a protocol used to transform the bacterium *Escherichia coli*. Read through this protocol and decide what the reasons are for including each stage. Then attempt SAQ 4.2. This will give you some experience of interpreting practical protocols as well as appreciating the steps that are necessary in the practical application of transformation.

In the experiment we are using an *E. coli* strain which is sensitive to an antibiotic and transforming DNA, in the form of a plasmid, carrying an antibiotic resistance gene.

Day 1.

Set up an overnight culture of *E. coli*.

Day 2.

1)   Inoculate 20 ml of nutrient broth with 0.75 ml of the overnight *E. coli* culture.

2)   Incubate, with aeration, for 90-100 minutes at 37°C (to an O.D.~0.6).

3)   Spin down the cells at 10,000 x g for 30 seconds.

4)   Wash the cells in 5 ml of 5 mmol l^{-1} NaCl and spin down at 10,000 xg for 5 minutes at 4°C.

5)   Re-suspend the cells in 2 ml of 100 mmol l^{-1} $CaCl_2$ and incubate on ice for at least 20 minutes.

6)   Spin the cells down at 10,000 x g for 5 minutes at 4°C.

7)   Re-suspend the cells in 1ml of 100 mmol l^{-1} $CaCl_2$.

8)   Add 25 μl of DNA to 200 μl of cells; incubate on ice for at least 1 hour; heat at 42°C for 2 minutes. Return to ice.

9)   Add 5 ml of nutrient broth and incubate with aeration at 37°C for 2 hours.

10   Plate out the transformants onto selective and non-selective media in Petri dishes in a top overlay.

Day 3.

Examine the Petri dishes.

SAQ 4.2

1) Why are the cells grown to an optical density of 0.6?

2) How are the cells calcium shocked?

3) What is the heat shock treatment?

4) What are 'selective media'?

5) What would be the expected results of this experiment on Day 3?

4.2.4 An example of choosing an appropriate transformation technique

For many, probably most organisms, we have no choice at all! There may only be one, or variations of one, basic technique available. However, as we described earlier, in some cases a number of methods may be used, although these may differ in their efficiencies. We will illustrate this using the yeast, *Saccharomyces cerevisiae*. The techniques available are:

- **Lithium acetate transformation.** Vigorously growing cells are 'shocked' with 100 mol l^{-1} lithium acetate, the DNA is added, followed by PEG and a heat shock is applied. A relatively fast and simple technique. Efficiency = 10^3 colonies per μg of DNA.

- **Protoplast transformation.** Protoplasts, produced using an enzyme (for example a glucanase) suspended in an osmotic buffer, are exposed to DNA with PEG and a heat shock is applied. A more difficult and lengthy method. Efficiency = $2\text{-}5 \times 10^4$ colonies per μg of DNA.

- **Electroporation.** Actively growing cells are pulsed at high voltage (1.5 kV) in the presence of DNA. A relatively quick technique which requires the use of specialised apparatus. Efficiency = 2.5×10^5 colonies per μg of DNA.

Do remember though that the efficiency of these techniques does very much depend on the individual worker! For efficient transformation by electroporation, it seems important that an osmotic buffer like sorbitol is used. Even though the cell walls are still present, presumably the membranes are made highly permeable and fragile during the electroporation process.

∏ Rank the above methods in order of efficiency. Above 100 ngs of DNA the efficiency of transformation declines for electroporation. Which other method would you choose?

You should have ranked them in the following order: 1) electroporation; 2) protoplast transformation; 3) lithium acetate. It would appear, therefore, that electroporation would be the best method. The protoplast method would be the next best bet although the lithium acetate technique would be simpler. Your final choice will be governed by the number of transformants needed.

4.3 Transforming DNA - survival in the new host cell

4.3.1 Basic principles

cloning and
amplification

If we are to use transformation DNA for DNA recombination *in vitro*, then we must be able to insert the DNA into a suitable host (that is, one in which it can be replicated). To achieve, this the DNA of interest has to be inserted into a suitable cloning vector. By using a cloning vector we can successfully produce multiple copies of the DNA of interest. We can describe such multiple copies as clones and the DNA of interest can be described as being amplified (cloned).

A cloning vector is a DNA molecule that will accept an insert of another DNA molecule thus allowing:

* replication of the insert ie by attachment to a suitable replicon (a DNA molecule with its own origin of replication);

* amplification of the insert to generate multiple copies;

* selection of the transformed cells within a population of potential host cells.

plasmids and
phages as
vectors

Most suitable vectors are plasmids or viral (usually bacteriophage) DNA as they are natural replicons. Other commonly used vectors are cosmids and yeast artificial chromosomes (YACs). We will examine these in more detail later. You should, however, realise that the cloning vectors that are now used are highly modified derivatives of natural vectors. They have been produced by a variety of techniques including restriction enzyme technology and DNA ligation. Those modifications have been designed to make them more amenable for use in the laboratory. For example, easily selected marker genes have been inserted and many carry restriction enzyme sites placed in strategic positions within their DNA sequence. This latter feature enables us to cut these vectors in predictable positions and to insert the desired piece of DNA. We will see some examples of these derived vectors in the following sections.

You should note that the discussion of cloning vectors provided in these sections is based on the information summarised in Chapter 3. We have included several reminders of the key issues. However, if you feel you need more information about plasmids and phages, we recommend the BIOTOL text, 'Genome Management in Prokaryotes' where the physical and biological properties of these entities are described in fuller detail.

4.3.2 Plasmids as cloning vehicles

Plasmids have many features which make them suitable candidates as cloning vehicles.

∏ Before reading on see if you can list these features.

You may have thought of several features. The main ones we hope you included are as follows. Plasmids are:

* replicons which are stably inherited separately from the chromosome;

- all genetically alike molecules for a specific plasmid type with a constant characteristic unit size;

- able to replicate independently of the chromosome;

- carriers for a limited number of genes, eg antibiotic resistance, transfer or mating functions, sugar fermentation or pigment production.

The ability to confer antibiotic resistance upon a host cell forms the commonest means of detection and selection of plasmids within a population of cells.

4.3.3 Desirable characteristics of plasmids

Π From the discussion given in earlier chapters, you should already have some idea of the characteristics we should look for in a plasmid which makes it a suitable candidate as a cloning vehicle. See if you can write down two or three features that are desirable in a cloning vector.

The sort of features you should have included are:

- cloning sites present (ie single sites for a large number of restriction enzymes). This will allow us to insert fragments produced by a range of restriction enzymes, using DNA ligase enzyme, to join the molecules together;

- cloning sites in genes with readily selectable phenotypes to allow recognition of vectors with inserts, ie having lost that phenotype because the cloned DNA insert has disrupted the plasmid gene;

- low molecular weight to allow manipulation of the molecule without shearing. Also, smaller plasmids are often found at a higher copy number (ie more per cell) than larger plasmids.

We must remember that, although a specific natural plasmid may not have all these desirable features, we can use recombinant DNA techniques to modify the molecule.

4.3.4 Sources of plasmids

Almost all of our original plasmid vectors are derived from the bacterium *E. coli*. However, a range of bacteria provide us with useful plasmids which can themselves be used as vectors or carry genes that we can use as selectable markers when cloned into other plasmids. Some eukaryotes, particularly yeasts, also provide potential plasmid vectors.

bacteria and yeast as sources of plasmids

The host range of many plasmids can be quite small and as we have an enormous stock of *E. coli* mutants and understand its growth and physiology better than that of any other organism, most manipulations rely upon the eventual use of *E. coli* strains and vectors.

4.3.5 Plasmid form

We remind you that plasmids normally exist as double-stranded, circular DNA molecules. However, many different forms are possible and these all find use in recombinant DNA technology. We illustrated some of these in Figure 3.4. Here we will illustrate them again but in simplified form (Figure 4.1). In this case the DNA is shown as two parallel strands. Remember, however, that these two strands, in nature, form an

topoisomeric forms of plasmids

α-helix. For our purposes here, it is easier to illustrate the DNA as shown in Figure 4.1. The various forms (topoisomers) of plasmid DNA are:

- covalently closed circles, CCC DNA; both strands of the molecule are intact;

- open circles, OC DNA; one strand of the molecule is cut;

- supercoils, SC DNA; the DNA molecule is further twisted;

- linear, L DNA; here both strands of the DNA have been cut.

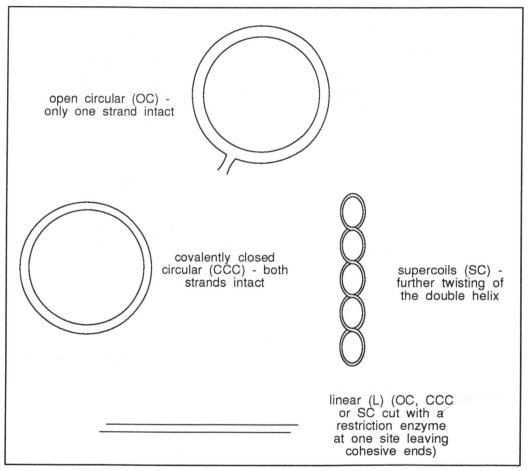

Figure 4.1 Topoisomers of a plasmid. *In vivo* these structural forms are interconverted by gyrase, topoisomerase, ligase and endonuclease enzymes. Supercoils run faster than the other forms on agarose gels.

We have also learnt that, because of their different configurations, OC, CCC and SC DNA can be separated on an agarose gel. Rare plasmid forms exist naturally as linear forms, for example, as in the giant plasmids of *Streptomyces*.

plasmid incompatibility groups It is also important to remember that plasmid incompatibility groups exist. Two plasmids from the same incompatibility group cannot co-exist in the same cell. Thus there are some restrictions on the number and variety of plasmids we can introduce into particular cells. The introduction of two plasmids from the same incompatibility group will lead to one or both of the plasmids being excluded. In *E. coli*, there are over 30 different incompatibility groups.

Π How can OC, CCC and SC DNA be converted to an intact L DNA form?

To achieve this, we need to incubate the DNA with a restriction enzyme which cleaves the DNA at a single restriction enzyme site.

polymeric forms of plasmids

Many plasmid molecules have the ability to associate together, usually as even multiples of the basic molecule. So, dimers are associates of two molecules and tetramers are 4 plasmids together. On an agarose gel, these 'higher forms' will run as DNA bands with apparent molecular weights equivalent to twice or four times the basic unit size. Once again, complete digestion of any of these higher forms with a restriction enzyme recognising a single site, will produce linear DNA.

4.3.6 Bacteriophage vectors

Lambda (λ), being a virus of *E. coli*, was extensively studied from the early days of molecular biology. It is a self-replicating DNA molecule within a protein coat. Several genes were already mapped by classical genetics and, by 1982, its entire DNA sequence was known.

cI repressor and the lytic cycle

Wild-type λ is a temperate phage. It can either enter a lytic life cycle or it can integrate into the host genome and so enter a lysogenic life cycle. In the lytic life cycle, a repressor gene (cI) is 'switched off'. In the lysogenic life cycle, this gene is switched on and the cI repressor prevents phage development.

Π If a bacterium carrying a lysogenic phage λ (as a prophage) in its genome is exposed to UV light, the cI repressor is inactivated. What will be the result of this exposure?

You should have concluded that the phage will enter a lytic cycle as the cI repressor would no longer prevent phage development.

cos sites and circularisation of λ DNA

DNA of phage λ is a linear duplex of about 48.5 kb although for much of its natural life cycle it forms a circular duplex. At the ends of the linear form are short single-stranded regions, complementary in sequence, which allow formation of the circular form. In other words, the linear form has naturally cohesive ends forming a *cos* site. This short sequence is essential for recognition of unit lengths of the λ genome. We shall see later how important the *cos* site is when packaging DNA into the virus head.

Functionally related genes tend to be clustered together on the map or chromosome. It is possible to insert foreign DNA into specific restriction sites and sometimes foreign genes can be expressed using the viral promoters.

A map of the major genes of wild-type λ is given in Figure 4.2. Note that some genes are associated with DNA synthesis and host lysis. Others (A, D, and E) are concerned with phage construction. cI, remember, is concerned with the regulation of phage development. It also bestows some immunity to infection of the host cell by other phages. For example, if a bacterial cell carries a lysogenic phage then there will be cI repressor molecules present in the cell. These will repress the replication of other phage particles capable of binding the same repressor. Clearly such phages are members of the same incompatibility group. Note also that on the wild-type genome there are segments which are involved in the integration of the phage genome into the hosts's DNA.

Similarly to natural plasmid vectors, natural or wild type λ DNA has the disadvantage of showing a number of restriction sites for any useful enzymes. All of our vectors are derivatives based on mutants which have lost particular restriction sites.

4.3.7 Derivative vectors

derivative vectors

Wild-type λ has been modified to produce two broad categories of derivative vectors:

insertional vectors

- insertional vectors have single cloning sites for the insertion of foreign DNA, eg λ gt 10;

replacement vectors

'stuffer' fragment

- replacement vectors have a pair of sites which allow a λ DNA sequence to be removed and replaced by foreign DNA, eg λ EMBL4. Much of the central region of the λ map is dispensable in this way and this increases the insert size that can be maintained. This central, dispensable region is known as the 'stuffer' fragment.

These vectors are illustrated in Figure 4.2.

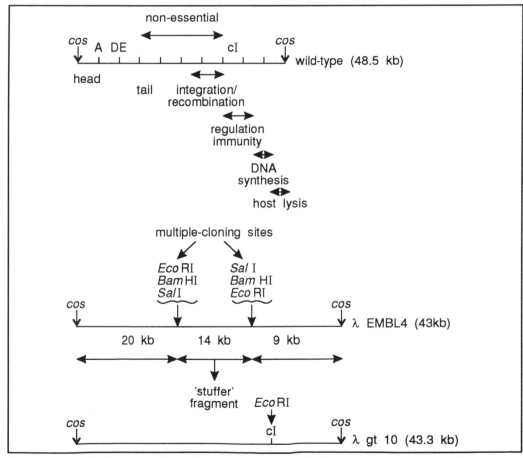

Figure 4.2 Bacteriophage λ and derivative vectors. (A) gene codes for an endonuclease which cleaves the phage DNA at specific sites (*cos* sites); (E) codes for the coat protein of the phage head; (D) codes for the product involved in the maturation of the phage (see text for details). Note that the *cos* sites are short, single-stranded regions. They are complementary in sequence and allow the DNA to be circularised.

The wild-type map shows a clustering of genes of related function required for the normal life cycle. As we have seen, λ EMBL4 is a replacement vector: the non-essential region (stuffer fragment) can be removed using the multiple cloning sites, allowing compatible inserts of 9-23kb. λ gt10 is an insertional vector allowing compatible inserts up to 7.6 kb into the *Eco* RI cloning site, inactivating the cI repressor. Remember cI repressor prevents lytic phage development. Insertion of DNA in this repressor site inactivates it, and as a consequence the phage on invasion into certain selective *E. coli* strains will cause lysis (ie form plaques). This allows us to recognise 'recombinant' vectors, (ie those carrying inserts of foreign DNA).

4.3.8 Cosmid vectors

concatamers

When λ DNA replicates it produces concatamers made of linear, repeating genome units. Concatamers (chains of λ DNA) can be packaged neatly into single chromosome lengths during the natural process of DNA replication in the host cell. For this to occur the system needs only to recognise *cos* sites that are 38 to 52 kb apart and any intervening DNA sequence will be packaged into infective phage particles (Figure 4.3).

cos sites

Such a particle can then infect suitable bacterial host cells. If the DNA molecule carries no phage functions other than *cos*, then no host cell lysis or plaque formation will occur. However, if the infective DNA carries other selectable markers such as antibiotic resistance, these characters may be conferred upon the host cell. Such vectors are

cosmids

termed cosmids.

capacity of cosmid vectors

We can think of a cosmid as being the ultimate replacement vector where almost all the viral genes have been deleted. The great advantage of cosmids is thus their large capacity for foreign DNA, although they are no longer true phage - infection of *E. coli* could never result in cell lysis or plaque formation.

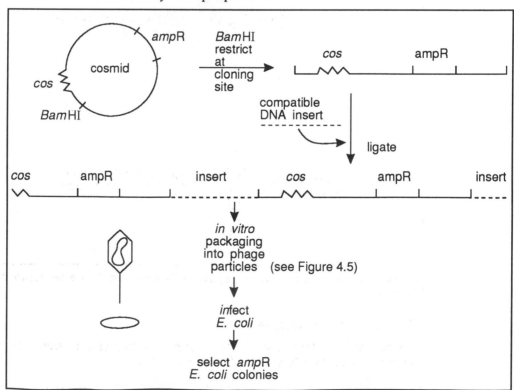

Figure 4.3 A simplified scheme for cloning in a cosmid. Ligation of linearised cosmid plus insert DNA produces concatamers with *cos* sites spaced at the correct distances for the packaging reaction to produce phage particles. Infection of *E. coli* produces antibiotic resistant cells.

4.3.9 *In vivo* λ **packaging**

During the lytic cycle the DNA of the virus is sorted into genome-sized chunks in a characteristic 'packaging reaction'. Look at Figure 4.4 to see how this works.

- The phage chromosome is replicated many times over within the host to form concatamers or chains of individual chromosomes linked together.

- Endonucleolytic cleavage by the product of gene A at *cos* sites allows accurate packaging of individual genomes into the pre-formed phage head. These pre-formed heads are made up of proteins encoded by gene E and eventually incorporate the gene D product as well.

- Tail proteins are added to produce a mature virus particle capable of infecting a new *E. coli* cell.

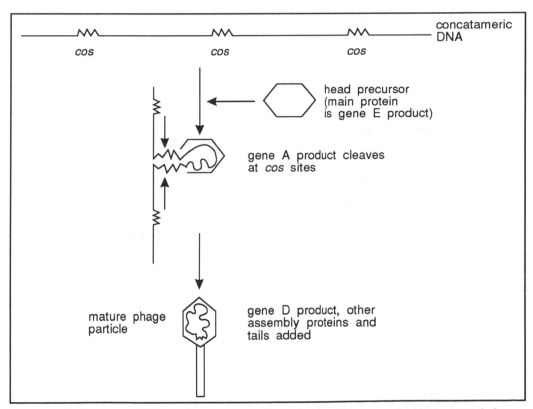

Figure 4.4 Packaging of λ DNA into phage particles. (Genes A, D and E are described in the legend of Figure 4.2).

4.3.10 *In vitro* λ **packaging**

Cloning foreign DNA into λ relies upon us being able to perform this packaging reaction *in vitro*. This is done in the following way:

The phage vector with its insert of DNA, is mixed with a very concentrated mixed lysate of two induced lysogens. By induced lysogens, we mean lysogenic phages that have been induced to enter a lytic cycle by exposure to UV light or by some other inducing agent. We use two mutant lysogens:

amber
mutations
- one blocked at the pre-head stage by an 'amber' mutation in gene D, so the precursors accumulate. Amber mutations produce codons in the DNA/RNA that prematurely end translation into protein;

- the other blocked from forming any head structure by an 'amber' mutation in gene E.

Thus if we mix lysates from the two lysogenic strains, we have all of the components we need to re-construct phages. If we also include suitable sized pieces of DNA these will be packaged into the phage heads. If we produce pieces of DNA containing *cos* sites (either cosmids or λ vectors) and other inserted DNA, we could in principle use the system to cut up the DNA into segments (using gene A products) and insert them into the phage (Figure 4.5).

∏ Examine Figure 4.5 and think how far apart *cos* sites have to be placed for the packaging reaction to work. Remember that wild-type λ has a linear piece of DNA of about 48.5 kb.

You probably concluded that you would need the *cos* sites about 48.5 kb apart in order to produce fragments of DNA of the right size to package. You would have been right but in practice we can use slightly smaller or larger fragments! Generally, the *cos* sites must be 38-52 kb apart, including λ genes and any foreign DNA insert(s). Write this distance onto Figure 4.5

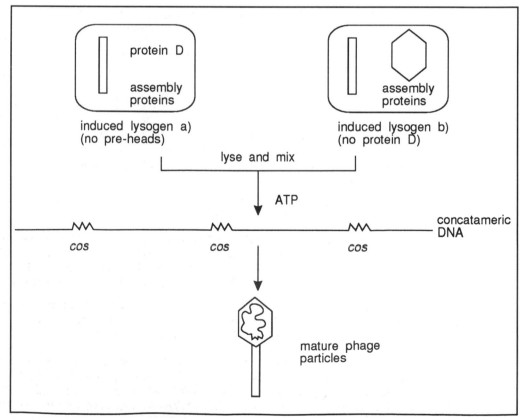

Figure 4.5 *In vitro* packaging of phage λDNA. Neither induced lysogen alone can package DNA but the mixed lysate has all the necessary components once the DNA is added.

| SAQ 4.3 | You have ligated a number of DNA fragments of various sizes into the *Eco* RI cloning site of λ gt10. Using a commercial packaging extract you produce a packaged suspension of total volume = 1 ml. Using 10 µl of this suspension you infect *E. coli* and produce 10 000 plaques in total when plated. If you initially used 0.5 µg of vector DNA, what is the apparent efficiency of the reaction in plaque forming units per µg of DNA (pfu per µg)? |

| SAQ 4.4 | Using the data presented in SAQ 4.3, you have calculated the efficiency of plaque forming units per µg DNA. You contact the manufacturer of the packaging extract and ask for the specified efficiency of the product. You are upset to find that the efficiency of the extract is quoted to be 1×10^9 pfu per µg DNA. Is there a logical explanation for the apparently disappointing result of your experiment? Can you get your money back from the manufacturers? |

4.3.11 Derivation of phage cloning vectors

multiple restriction sites of wild-type λ removed

As wild-type λ has far too many sites for most restriction enzymes in common use, before it could be adopted as a cloning vector some of these sites had to be removed. We could not practically try to insert foreign DNA into wild-type λ as cutting with most enzymes would cause the molecule to produce a large number of fragments that would be almost impossible to put back together again. This removal of restriction sites was achieved by a combination of classical *in vivo* techniques and *in vitro* methods. The sequence was as follows. (Figure 4.6 shows the basic scheme).

- Mutant phages were selected which had deletions resulting in the loss of two *Eco* RI sites at the left-hand side of the DNA molecule.

- From these phages, further mutants were selected which lacked the remaining *Eco* RI sites. These mutants were selected in the following way. Phage were alternately infected into *E. coli* hosts which contained or lacked the *Eco* RI restriction-modification system. Only phage which lacked *Eco* RI sites could survive in a host which produced *Eco* RI as a defence mechanism against infecting phage. Eventually, from the number of plaques formed from known numbers of phage particles in *Eco* R$^+$ and *Eco* R$^-$ hosts, it was demonstrated that phage were produced that completely lacked *Eco* RI sites.

- These *Eco* RI site mutants were then crossed, in a mixed infection of *E. coli*, with phage which still had intact *Eco* RI sites.

- Recombinant phages were then produced, some of which contained only a limited number of sites for *Eco* RI. Some of these phages had *Eco* RI sites conveniently positioned to allow the removal of the central, non-essential region by digestion with *Eco* RI. Hence, these phage could then function as replacement vectors.

Using the same idea of the selection of mutant phages with missing restriction sites and recombination with other phages, derivatives with reduced or absent sites for other restriction enzymes, like *Hin* dIII, were also produced.

There is now an enormous range of λ derivative vectors available. In some cases it is possible to ligate insert DNA directly in frame with λ promoters. One interesting and useful development has been the construction of vectors with improved biological containment properties. Phage Charon 3A, for example, has amber mutations in genes A and B. Such a phage will only produce infective virus particles on specific strains of

biological containment properites

E. coli and not on wild-type *E. coli*. The chances of such a vector 'escaping' from the laboratory to infect wild-type *E. coli* in humans are thus remote.

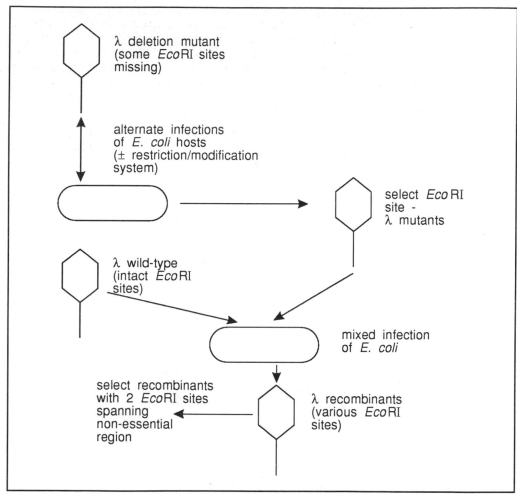

Figure 4.6 A scheme for the derivation of λ cloning vectors.

4.3.12 Isolation of vector DNA

Before we can consider cloning target foreign DNA into a vector we must first obtain a pure and concentrated preparation of the vector DNA itself. This DNA must be 'clean' enough to be cut with restriction enzymes and concentrated enough to allow us to transform organisms which have low efficiencies of uptake of DNA. The vector DNA should be essentially free of other contaminating molecules such as chromosomal DNA, proteins, oligosaccharides, RNA and DNA degrading enzymes.

The life cycle of bacteriophage λ provides us with an obvious opportunity to separate the vector DNA from the host cell. If a liquid culture of *E. coli* cells is synchronously infected with a large number of phages, then after a short period of time, virus particles will be released from the lysing host. The suspension is then centrifuged briefly to precipitate cell debris leaving the phages in suspension. It is then a simple matter to precipitate the phages themselves using polyethylene glycol (PEG). Finally, a phenol

vector DNA prepared from phages is released during lysis

extraction will remove the phage protein coats and the purified DNA can be recovered by precipitation with ethanol or isopropanol.

Many methods are available to separate pure plasmid DNA from its host bacteria. We described some of these in Chapter 3. You should remind yourself of these procedures. The most common method makes use of the differences in buoyant density of chromosomal and covalently closed circular DNA in the presence of ethidium bromide (EtBr) using isopycnic centrifugation and a caesium chloride gradient (Figure 4.7).

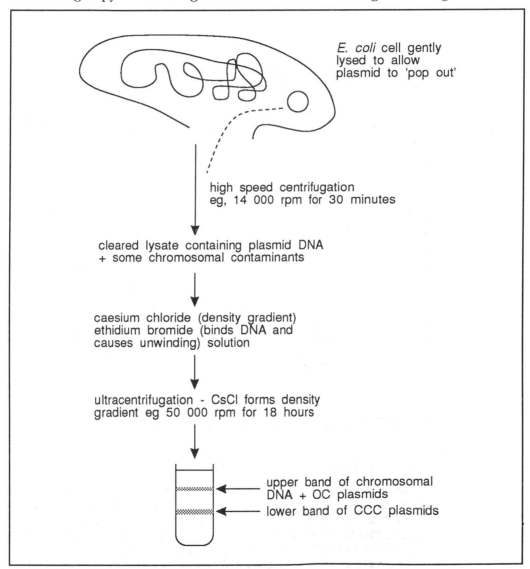

Figure 4.7 Isopycnic centrifugation on a caesium chloride gradient to isolate plasmid DNA.

We remind you that CCC DNA can only unwind to a limited extent, less EtBr is bound, and CCC DNA then has a higher density than chromosomal DNA. The band of DNA is removed using a syringe needle and caesium chloride/ethidium bromide is subsequently removed by dialysis/isopropanol extraction to yield pure plasmid DNA.

SAQ 4.5

We can create new vectors by inserting and deleting sequences at existing restriction sites. Natural processes of recombination do this all the time in bacterial populations. For example, in *Staphylococcus aureus*, multiple drug resistance plasmids carry resistance to a number of antibiotics and cause terrible infection problems in hospitals.

Using your own knowledge of cloning put the following steps in sequence to create a new vector from the two plasmids (A and B) illustrated in Figure 4.8.

1) ligate fragments;

2) isolate fragments of correct size from agarose gel;

3) digest DNA with *Eco* RI and *Pst* I to produce two restriction fragments/plasmids;

4) isolate plasmid DNA from parental host cells;

5) select transformants for ampR (ampicillin resistance) and tetR (tetracycline resistance);

6) transform *E. coli*.

Using your answer to SAQ 4.5 insert labels for the key steps in Figure 4.8

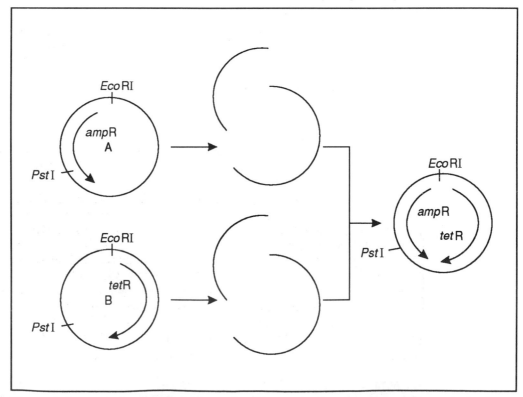

Figure 4.8 Hypothetical construction of a new plasmid vector. In this case we have a strong selection for our new vector (*amp* R and *tet* R) and isolation of the fragments from gels is not really necessary. However, in many cloning experiments the only clue to the correct fragment to sub-clone may be its size. Arrows inside circles denote the direction of the transcription of the genes.

4.3.13 Monitoring the cloning experiment

How can we be sure that our cloning experiment has worked? In reality no single enzyme reaction is 100% efficient. At each stage, we run samples on mini-agarose gels to assess the efficiency of linearising the plasmid or of the ligation reactions. The size and number of bands on the gel will indicate the success or otherwise of the reactions.

Often a key problem is that when we mix the linearised vector and the insert with the ligase and buffer the plasmid will recircularise on its own without an insert. Two common methods are used to combat this:

- performing the reaction at high DNA concentrations will favour inter- rather than intramolecular reactions;

- pre-treatment of linearised vector with calf intestinal alkaline phosphatase (CIP) will remove 5' phosphate groups. Such treated vectors cannot recircularise on their own and require an intact phosphorylated insert to complete the ligation.

After a typical ligation reaction, we can expect to find a range of DNA molecules. In addition to recircularised vectors alone, we will also find unligated linear vectors and insert fragments. More complex, larger DNA molecules could result from ligation of more than one vector and/or insert. From such a mixture, we hope to find single vector molecules with single inserts - the object of our cloning operation!

Only circularised vectors will transform *E. coli* efficiently and allow for selection with the appropriate markers, for example ampR or tetR. Larger plasmid constructs are at a disadvantage to smaller molecules in transformation.

Finally, we can physically check for the presence of plasmids with inserts of the correct size. We can perform simple, rapid alkaline lysis of small samples of transformed cells and run the products on agarose gels along with suitable markers of known size.

4.3.14 Selection of recombinant DNA molecules

We have already seen that by increasing the DNA concentration in a ligation reaction and/or by the use of CIP, we can increase the chances of producing a recombinant molecule. When cloning in phage or cosmid vectors, the λ packaging reaction itself acts as a recombinant selection process. Sequences 38-52 kb between two *cos* sites will be packaged. So cosmid vectors alone or λ replacement vector 'arms' alone will be too small.

use of λ gt10 to select recombinant DNA molecules

Additionally, with some phage vectors, we can utilise specific *E. coli* mutant hosts to select for vectors with inserts. We met with an example when we were discussing phage λ vectors (Section 4.3.7). You should recall that the vector λ gt10 has an *Eco* RI site adjacent to the cI repressor gene. Inserting DNA into this *Eco* RI site inactivates the repressor gene. Thus λ gt10 produces cI repressor and the λ gt10 enters the lysogenic life-cycle when infected into suitable *E. coli* cells. On the other hand λ gt10 containing an inserted DNA sequence does not produce cI repressor and the phage vector enters the lytic-cycle. Thus, using this vector we can recognise recombinants by their ability to form plaques. Usually, strains of *E. coli* (for example, *hfl*A) which show a high frequency of lysogeny are used. We can present this system in the following way:

A	non-recombinant	→	hflA strain	→	no plaques	lysogenic
B	recombinant (cI repressor inactivated)	→	hflA strain	→	plaques	lytic

*hfl*A = high frequency of lysogeny, *E. coli* mutant.

Each plaque formed on the *hfl*A strain would represent a single cloned DNA molecule amplified many times over during phage infection and replication. We could then be sure that the ligation and packaging reactions had worked.

4.3.15 Shuttle vectors

Plasmids vary in their natural host range - some are very host specific whilst others have a broad host range amongst bacteria. We can use the 'natural' markers of antibiotic resistance, for example, to select for successful transformation of natural hosts. However, if we wish to attempt to transform a second organism with such a vector we would need to introduce a second selectable marker for the alternative host.

Vectors with two selectable markers that allow the transformation of two different organisms are termed shuttle vectors. These vectors are the key to any inter-species cloning strategies. Additionally, they allow us to amplify small amounts of DNA in *E. coli* when transformation frequencies are low for the alternative host. Large quantities of DNA can then be prepared from *E. coli* for introducing into the new host.

shuttle vectors have two sets of selectable markers An example of an *Aspergillus nidulans* (fungus) *E. coli* shuttle vector is pDJB3 which carries a normal *pyr*G$^+$ allele to complement the inability of the *pyr*G$^-$ *A. nidulans* host to grow in the absence of pyrimidine supplements. The selectable marker for *E. coli* transformation is ampR (ampicillin resistance).

Whilst the shuttle vectors that we use are all artificial constructs, it may be that similar inter-species gene transfer events occur in nature to give rise to 'horizontal transfer' of genes between distantly related organisms. For example, the genes that code for β-lactam antibiotic synthesis in streptomycete bacteria and fungi like *Penicillium* all show remarkably similar sequences. We think that the fungi may well have obtained these genes directly from the bacterial host possibly by the action of a 'natural' shuttle vector.

4.3.16 Yeast artificial chromosomes

many genes are large For some purposes it is necessary for us to clone very much larger inserts of DNA than the vectors so far considered will allow. Many human genes for example, are extremely long; the Duchenne muscular dystrophy gene is greater than 1 mega base (Mb) in size (1Mb = 1000 kb). Additionally, when we wish to screen a large genome, like the human genome, for specific genes, we would have to make a ridiculously large number of constructs in traditional vectors like plasmids, phages or even cosmids, (we will discuss this further in the next chapter). Clearly, there is a need for cloning vectors with a very large insert capacity.

In a sense, we can consider normal eukaryotic chromosomes as being cloning vectors of enormous size, which stably replicate a huge number of genes. For our purposes we can think of the following three key features of eukaryotic chromosomes:

* they contain a centromere region;

* they have telomere regions at each end;

* they have multiple origins of replication.

These three features are essential for the following reasons. The centromeric region is required in order for the chromosome to become attached to the mitotic spindle during cell division. Without such a centromeric region, the daughter chromosomes would not become properly segregated during mitosis. The telomere regions of chromosomes are essential as they are constructed in such a manner to ensure that the ends of the DNA double helix located in these regions are completely replicated. Although the molecular details of the processes involved at the telomeres at the tips of each chromosome are not fully understood, chromosomes without telomeres are not properly replicated and are eventually lost. Finally, multiple origins of replication are required in order to complete DNA replication in a reasonable period of time.

YACs Yeast artificial chromosomes (YACs) have now been made that incorporate these features of natural chromosomes. They contain:

- a centromere region and an 'autonomously replicating sequence' (*ars*) which enhance the stability of the construct, especially with large inserts of DNA;

- yeast telomeres of irregularly repeating units (5' CCCA 3') of about 100 bp in length. These preserve the ends of chromosomes by ensuring the completion of DNA replication;

- a yeast selectable marker is incorporated on each 'arm' to identify intact constructs.

A simplified version of a yeast artificial chromosome is illustrated in Figure 4.9. Examine this figure carefully. Normally the YAC vector is maintained as a circular molecule until it is needed. Then it is cut in at least two places.

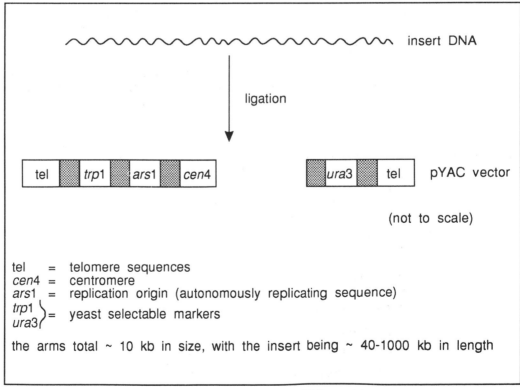

Figure 4.9 Construction of a yeast artificial chromosome. The pYAC vector is maintained as a circular molecule before being cleaved by restriction enzymes at two or more sites to produce the two 'arms'.

SAQ 4.6

Factor VIII is one of at least 10 proteins required for normal blood clotting. A mutation in the factor VIII gene leads to the sex-linked trait of haemophilia A.

The human X chromosome has been estimated to be around 200 Mb in size. If the gene for factor VIII represents 0.1% of this chromosome which of the following vectors could be used to clone the intact factor VIII gene?

1) a plasmid of maximum insert size of 10 kb;

2) a cosmid of maximum insert size of 40 kb;

3) a YAC of maximum insert size of 1000 kb.

SAQ 4.7

The following is a description of the components of a vector called pCAP2.

1) *cos* - packaging recognition site.

2) *ans*l - *Aspergillus nidulans* sequence 1, improves *A. nidulans* transformation frequency.

3) *pyr4* - confers on *A. nidulans* transformants the ability to grow in the absence of pyrimidine supplements.

4) ampR - ampicillin resistance.

5) *ori* - origin of replication.

6) *Bam* HI - cloning site.

7) Total size = 8.65 kb.

This vector is illustrated in Figure 4.10.

Which of these terms describes pCAP2?

a) phage; b) cosmid; c) shuttle vector; d) YAC; e) plasmid.

You may choose more than one. Describe your reasons.

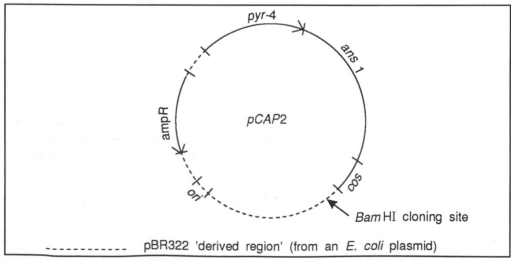

Figure 4.10 Vector described in SAQ 4.7. This molecule carries a number of important sequences that determine its role as a cloning vector. There are multiple sites for many restriction enzymes, so most restriction maps of vectors show only those important for cloning.

<table>
<tr><td>

SAQ 4.8

</td><td>

1) What would be the approximate maximum and minimum insert sizes for pCAP2? (See SAQ 4.7).

2) How would pCAP2's value as a cloning vector be diminished if we deleted *pyr*4 and *ansl*?

</td></tr>
</table>

4.3.17 Single-stranded (SS) DNA vectors

advantages of single-stranded vectors

All of the vectors that we have considered so far consist of double-stranded DNA molecules. However, some filamentous phages like M13 have a single-stranded, circular DNA molecule as their 'chromosome'. Such a molecule has a number of advantages over the double-stranded (DS) cloning vectors. These advantages are:

• DNA sequencing by the dideoxy method, some protocols for making probes and *in vitro* mutagenesis require SS DNA;

• the phage can efficiently transfect competent *E. coli* cells to give either plaques or colonies depending on the selection method employed;

• there are no packaging constraints as the viral DNA size determines the phage particle size. Quantities up to 6 times the normal amount of DNA have been reported to be packaged.

The phage DNA is copied via a double-stranded, circular DNA intermediate, the so called RF or replicative form, which can be isolated and manipulated easily just like a plasmid.

Often a cloning strategy will use a plasmid or phage vector for screening for a particular gene. Then this could be sub-cloned into M13 or one of its derivatives like the 'Bluescript' vectors, for sequencing or making probes. Note that the dideoxy method for DNA sequencing is described in detail in the BIOTOL texts, 'Analysis of Amino Acids, Proteins and Nucleic Acids' and 'Genome Management in Prokaryotes'. We will discuss the construction of probes in more detail later. Essentially, a probe in this context means a single-stranded DNA molecule which can be used to 'identify' DNA molecules carrying complementary nucleotide sequences.

4.4 Choice of a cloning vector

You have now met examples of a range of cloning vectors. Some, like YACs, have quite specialised uses whilst others have a much broader set of applications. We can compare some of the strengths and weaknesses of three of the most common types of cloning systems.

Ultimately, the choice of a cloning vector will rest on a variety of factors among which are:

• the organism to be transformed;

• the selectable markers available;

• the screening procedure envisaged to screen for inserts of interest;

- the amount of genetic material we wish to transfer or clone;

- the numbers of transformants/transfectants required;

- the personal preference and expertise of the worker!

∏ You might like to attempt to work out the relative advantages and disadvantages of plasmid, phage and cosmid vectors and then compare them to those we described below. Alternatively, read through the remainder of this section and then construct yourself a summary table using the following headings:

Vector type	Advantage	Disadvantage

Plasmids

- Advantages; usually small and easy to handle with a large number of potential selectable markers for transformation.

- Disadvantages; will only take relatively small inserts (in practice 10-15 kb in size would be a common upper limit) with transformation efficiency never being as high as λ infection (per μg DNA). Recombinant plasmids with inserts can often be hard to select from recircularised plasmids alone.

Phages

- Advantages; can take larger inserts than plasmids (maximum being around 23 kb depending on the derivative type). Recombinants can be selected by packaging reactions and growth on selective bacterial strains. Easy to screen for specific genes using probes against plaques (using single-stranded DNA probes to identify plaques containing DNA with complementary DNA sequences). Highly efficient natural infection process for bacterial host.

- Disadvantages; often hard to transform or transfect organisms other than the natural host. Large inserts may become 'scrambled' by recombination, or more than one insert may be cloned into a single site.

Cosmids

- Advantages; take very large inserts (up to 45+ kb) thereby reducing the numbers of clones that need to be screened. Again, packaging imposes a size selection for recombinants and provides a highly efficient means of infecting host bacteria.

- Disadvantages; it is often technically difficult to isolate and restrict high molecular weight DNA to provide inserts of sufficient size. The same drawbacks as apply to λ can operate in both cosmids and phage. Due to the large inserts these can often become 'scrambled' by recombination and therefore we often have to carry out quite extensive further analysis to ensure we have cloned DNA sequences of interest within the cosmid.

Let us see how good you are at selecting suitable vectors by attempting SAQ 4.9.

It should be self-evident that there is often a choice of more than one cloning vector that may be used. For a particular organism your final choice may be determined by the availability of a shuttle vector. But crucial to any choice of vector is the need to have a suitable screening system. The next chapter explains how important it is to consider the screening strategy before we choose a cloning vector.

SAQ 4.9	Choose the best type of vector(s) for the following cloning jobs.

Choose the best type of vector(s) for the following cloning jobs.

1) To sub clone a bacterial gene of 1.5 kb in length from a 'giant' plasmid of 105 kb.

2) To clone a human gene of length 800 kb intact.

3) To sub clone a sequence of length 15 kb into a vector that will produce 10^6 E. coli clones.

You may choose one or more from a)-d):

a) plasmid; b) cosmid; c) YAC; d) phage.

Summary and objectives

The main objective of this chapter has been to help you to appreciate the range of techniques available for the introduction of foreign DNA into host cells. We have emphasised that it is important to consider both the type of DNA molecule used in transformation and the physical process of getting it inside the cell. The main focus of the chapter has been on the properties and features of plasmid, phage, cosmid and yeast artificial chromosome vectors.

Now that you have completed this chapter you should be able to:

- calculate the efficiency of transformation from provided data;

- identify the purpose of particular stages in transformation protocols;

- make informed choices of techniques where more than one protocol for transformation is available;

- describe the natural origins and derivation of plasmid, phage, cosmid and YAC cloning vectors;

- distinguish between the major types of vector from descriptions of their genetic constitution;

- appreciate the value of *in vitro* λ and viral infection in phage and cosmid cloning strategies;

- choose the most appropriate type of vector for a specific cloning job.

Gene isolation

Gene isolation

5.1 Introduction

The power of modern molecular biology is centred upon our ability to isolate individual genes and to clone them. During the cloning procedure many identical copies of each single gene are made and this opens the way to both the study of the fine structure of the gene and gene manipulation.

A crucial problem in this type of work is that the gene of interest may represent only a tiny portion of the total amount of DNA isolated from an organism. We, therefore, have to look for a method that enabled amplification of specific parts of the genome. A general strategy for this is as follows:

* fragmentation of the organisms's DNA into pieces of the correct size to insert into a cloning vector;

* ligation of the DNA inserts into the cloning site of the cloning vector;

* introduction of the recombinant vector containing the DNA inserts into host cells;

* culture of host cells;

* screening of the cells for expression of the gene of interest.

The final clone selection process is a very crucial stage of any recombinant DNA experiment; it is often the most lengthy part as well. Therefore it is important to choose a cloning vector that will make clone selection easier.

The present chapter will deal with the techniques for selective gene isolation in a step-by-step way. First the construction and storage of genomic DNA banks and cDNA banks will be described and then the different methods for the screening of gene banks will be considered. We will see that during construction of DNA banks and their screening we often need specific short DNA sequences to use as probes. The final section of the chapter describes *in vitro* DNA synthesis.

5.2 Genome size and genomic DNA libraries

When we physically isolate total DNA from an organism, we obviously have a sample of all the genes of that organism. Our first problem is to assess the size of the task that faces us. How many other genes do we have to sift through to find the gene that we are interested in? Our clear starting point is to find out how much DNA is represented by a single genome and then estimate how many genes this may constitute.

Let us first think about the amount of DNA per genome by considering some examples in Table 5.1. If we think of a gene as a genetic unit of function or a DNA sequence specifying a particular, inherited trait, how many genes do the genomes in Table 5.1 represent?

Organism	Genome size (kb per haploid)
φX174 (virus)	5.4
E. coli (bacterium)	4×10^3
Saccharomyces cerevisiae (fungus)	1.4×10^4
human	2.8×10^6

Table 5.1 Approximate genome sizes of a range of organisms.

overlapping
genes in φX174

It is not easy to assign a number of genes to a particular genome. In the case of φX174 the tiny amount of DNA has forced this virus to adopt a highly unusual strategy, that of overlapping genes. The DNA sequence can be read in different reading frames using different start sites. Even so this single-stranded DNA virus only has room for about 9 different genes. At the other extreme the human genome is clearly enormous! However, not all of this DNA will represent genes that code for identifiable products - for much of this DNA the function is not known but it may be important in DNA processing or packaging. Nevertheless, in complex eukaryotes we are still talking about a large number of genes. A common estimate of the number of genes per eukaryote cell is around 50 000!

gene library
gene bank

Clearly, for all but the most simple organisms, to stand any chance of cloning a particular gene we must incorporate a very large number of DNA inserts into cloning vectors. Such a collection of recombinant vector/insert molecules which represents the entire genome of an organism is called a 'gene library' or 'gene bank'. A truly representative gene bank will contain recombinant molecules for every gene from that organism. We shall see shortly how to apply a simple equation to calculate the number of independent clones required for such a bank. Clearly, the more clones present, the greater the chances of cloning a specific gene.

5.2.1 Construction of genomic libraries

genomic DNA
bank

The simplest type of DNA library is a genomic DNA bank (Figure 5.1). The procedure for constructing it is as follows:

* extract 'total DNA' from the organism (see Chapter 3);
* cut the DNA with a suitable restriction enzyme;
* ligate the DNA insert into the cloning site of the vector.

The restriction enzymes we usually choose have 4-6 bp recognition sites occurring more-or-less at random in the DNA. We usually perform only a partial digest so that not all the restriction sites are cut and the DNA is not split into too small pieces. We also have to make sure that the cohesive ends of the DNA insert are compatible with those found in the cloning site of the vector.

Many vectors are suitable for producing genomic DNA banks, including plasmids, phages, cosmids and YACs. Chapter 4 outlines some of the factors that may determine your choice. Plasmid gene banks are often chosen for organisms with relatively small genomes, whereas, to limit the number of clones, YACs may be required for more complex and larger genomes. On the other hand, phage DNA banks provide us with easy methods for screening for specific genes via hybridisation. The cloning capacity of the chosen vector is critical, with plasmids being suitable for smaller inserts (say 10-15kb), cosmids taking inserts of 45+kb and YACs capable of cloning 1000 kb.

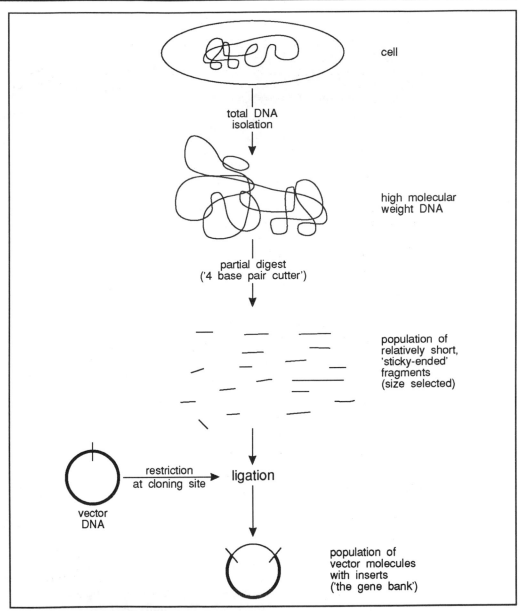

cell

total DNA
isolation

high molecular
weight DNA

partial digest
('4 base pair cutter')

population of
relatively short,
'sticky-ended'
fragments
(size selected)

restriction
at cloning site → ligation

vector
DNA

population of
vector molecules
with inserts
('the gene bank')

Figure 5.1 Generalised scheme for the construction of a genomic DNA bank. All the DNA molecules are double-stranded with the exception of short single-stranded overhangs at 'sticky ends'. Inserts of specific sizes may be selected using sucrose gradients or agarose gels.

SAQ 5.1

We want to produce a genomic DNA library. For producing the DNA inserts we apply a complete digest using an enzyme with a 6 bp recognition sequence like *Bam* HI. Describe the disadvantages of this strategy.

5.2.2 How many clones are needed for a genomic DNA bank?

This is really the next most obvious question. Clearly the number of recombinant molecules required in the library will be dependent upon two crucial variables:

- the genome size of the organism.

- the cloning capacity (insert size) of the vector;

We can make some assumptions and then derive a simple equation that will allow us to make a rough calculation of the size of the library we need for cloning a particular gene. The assumptions are:

- there is a complete and random representation of sequences;

- each fragment is of the same size;

- the genome size is known.

In practice none of these assumptions completely holds! However, they are accurate enough for practical purposes and we can use the following equation:

$$N = \frac{\ln (1 - p)}{\ln (1 - x/y)}$$

where, N = the number of clones needed to have a probability (p) of containing a given sequence. Usually we set p = 0.95 or 0.99 (ie to give a 95% or a 99% chance!): x = the insert size (in kb): y = the size of the haploid genome (in kb): ln = 'natural' logarithm.

It will be evident that organisms that are not haploid may require much larger libraries if specific heterozygous sequences are to be cloned. For example, a sea urchin has a haploid genome size of 8.6×10^5 kb and a genomic DNA bank of 8.8×10^4 clones would have a 0.99 probability of containing a given sequence using an insert size of 45 kb. An insert of this size would be typically suited for insertion in a cosmid, ie a vector containing plasmid sequences along with the phage λ *cos* site, allowing packaging into phage particles, (see Chapter 4).

| SAQ 5.2 |

If the fruit fly, *Drosophila melanogaster*, has a haploid genome size of 1.4×10^5 kb, how many clones would be needed in a genomic DNA bank to have a 99% chance of having a particular sequence in a cosmid of cloning capacity 45 kb?

5.3 cDNA banks

In the following sections we will provide you with details of how we construct cDNA banks and the techniques used to screen these gene banks.

5.3.1 Construction of cDNA banks

The second type of gene bank commonly used is a cDNA bank. cDNA stands for complementary DNA, by which is meant DNA molecules synthesised using the enzyme reverse transcriptase and mRNA as a template. Therefore, a cDNA bank will only represent the genes which are expressed (ie transcribed) in the cells of the starting material. This can be an advantage when we are working with eukaryotic cells in which a large part of the DNA is not coding for recognisable genes but has a regulatory or structural function. Of course, a genomic DNA bank would include this 'non-coding' DNA as well. Moreover, only a fraction of the total coding DNA may be expressed in a tissue or at a particular stage of development which may also determine the size of the cDNA bank.

Thus cDNA banks are usually prepared from eukaryotic organisms. Eukaryotic mRNA molecules carry a chain of adenine residues (the polyA tail) at one end and, therefore, can be separated from ribosomal RNA (rRNA) by capture on an oligo dT column. Prokaryotic mRNA does not have a polyA tail and cannot be purified in this way.

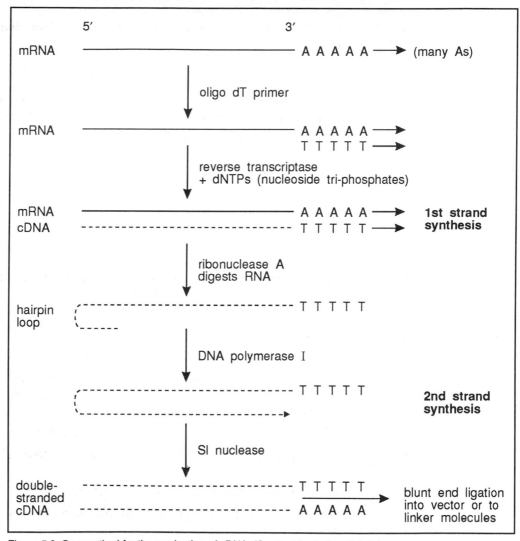

Figure 5.2 One method for the production of cDNA. (See text for further details).

Figure 5.2 outlines the procedure of constructing a cDNA bank. The following steps can be discriminated in a typical protocol:

• total RNA is isolated from the cells of the starting material;

• for most applications the mRNA is subsequently separated from the total RNA, for example by using an oligo dT column;

• in the presence of a specific buffer and RNase inhibitors, the enzyme reverse transcriptase synthesises complementary cDNA strands to the mRNA template. For this, the single-stranded mRNAs are 'primed' with short oligonucleotide sequences. These primers can be oligo dT which attach to the polyA tail, or other sequences attaching to complementary sequences of the mRNAs. In the presence of nucleotide triphosphates the enzyme reverse transcriptase then makes cDNA copies of the mRNAs obeying Watson and Crick base-pairing rules;

- after synthesis of the first DNA strands, the mRNAs are degraded by using the enzyme ribonuclease A;

- DNA polymerase I is used to synthesise double-stranded cDNA molecules;

- the hairpin loops are removed using S1 nuclease; this is a crucial step in the procedure at which DNA can be lost. In some methods, gaps are produced in the mRNA part of the cDNA/mRNA hybrids thereby allowing for more efficient priming of the synthesis of the second strand;

- the double-stranded cDNAs are ligated into the cloning site of a vector; often special linker or adaptor sequences are first added to the blunt ended cDNA to allow for ligation into a specific cloning site. Often cloning vectors are chosen in which the cDNA can be ligated in frame to a suitable promoter that will allow expression of the cDNA; λ gt11 is often chosen for this purpose.

∏ What macromolecule do we isolate from a cell in the first step of making: 1) a genomic DNA library; 2) a cDNA library?

We isolate 1) DNA (total) and 2) RNA (selecting for polyA mRNA in eukaryotes to avoid swamping with ribosomal sequences).

5.3.2 How many clones are needed for a cDNA bank?

It is much more difficult to estimate the number of clones required for a representative cDNA library than for a genomic DNA library. The frequency of a particular cDNA clone in the library will depend upon:

- the number of copies of the gene in the genome of the host cell;

- the level of transcription of this gene (and thus the relative amount of mRNA in the sample);

- the ease with which the mRNA can be copied into cDNA.

In theory we adopt the same statistical approach that we used for the genomic bank and we derive a similar equation:

$$N = \frac{\ln (1 - p)}{\ln (1 - n)}$$

where:

N = number of clones required;

p = probability of isolating the clone (usually $p = 0.95$ or 0.99);

n = fractional proportion of the total mRNA population represented by a single mRNA;

ln = 'natural' logarithm.

Estimates of the abundance of mRNA are notoriously difficult, so n is the most crucial factor here!

Of course, the cloning capacity of the vector and the genome size of the organism will also determine the size of the library that we need.

Let us suppose that a particular mRNA molecule is present in immature chicken red blood cells (RBCs) at around 10 copies per cell, which makes this quite a rare transcript. Let us assume further that the rare mRNA's make up together about 10% of the total mRNA of this cell type, and that there are only around 100 different mRNA molecules of this sort. So, the fractional representation of any mRNA of this rare group in the total mRNA population of the cell is around 1 in 1000 molecules (ie 10/10 000).

If $n = 10^{-3}$, using the above equation we only need to make a library of around 5000 clones to have a 99% chance of cloning our target cDNA.

However, in other cells types which have a greater variety of mRNA molecules we may expect to need a much larger number of clones. A typical mammalian cell may contain 20 000 different types of mRNA molecules and to clone a very rare mRNA may require a library of around 10^6 clones!

∏ Why should one cell type, like the immature chicken RBCs have fewer mRNAs than, for instance, the typical mammalian cell in the example above?

The solution hinges on the pattern of gene expression in different cell lines. As a general rule, differentiation of cells into different tissues does not result in the loss of genetic material (DNA) but stems from the preferential switching on and off of particular genes. Hence, depending on the metabolic state and the degree of differentiation, different cell types may produce varying types and numbers of mRNAs.

5.3.3 Enriching cDNA banks for specific clones

inducible genes

We have seen that cDNA libraries allow us to clone genes which are preferentially expressed in different cells. In some cases, this allows us to limit the number of clones that we have to screen. There are situations though, where we can further enrich our library for a particular sequence. We can do this by making use of the knowledge that some genes are inducible, ie their expression (level of transcription), can be increased by specific stimuli.

For example, we may be interested in the genes which are induced as part of a tolerance response when a plant is exposed to toxic quantities of heavy metals. If we 'shock treat' roots with the heavy metal, we may induce over-expression of these genes and so increase their fractional proportion in the mRNA pool. mRNA extracts of plants treated in this way will allow us to make a cDNA bank which is enriched for specific sequences.

5.3.4 Gene expression and cDNA banks

In many cloning experiments it is common to try to express a eukaryotic gene in a host like *E. coli*. However, when doing this we will be faced with a number of problems and we shall examine briefly a few important issues below.

Figure 5.3 shows a simplified diagram of eukaryotic gene expression. In what respect does this process differ from prokaryotic gene expression? One difference is that eukaryotic genes contain introns that are removed during RNA processing. There are also important differences in the promoter sequences. We will discuss these both in turn.

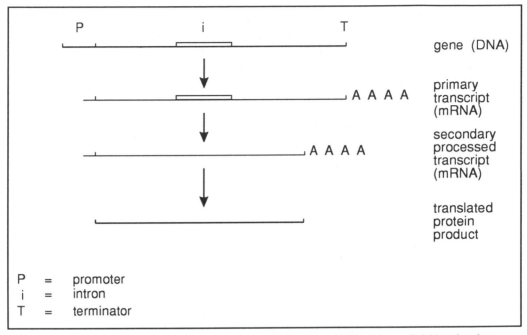

P   =   promoter
i   =   intron
T   =   terminator

Figure 5.3 Simplified model of eukaryotic gene expression. RNA polymerase binding is initiated at the promoter. Terminator sequences add a polyA tail (of 100-200 A's) at the 3′ end of the mRNA. The intron is spliced before translation into protein. The mRNA carries a 'cap' of 7 methylguanosine at the 5′ end to direct the ribosome to start translation at the start (AUG) codon.

Promoter sequences

Prokaryotes typically have the following consensus sequences within their promoters which act as recognition/binding sites for RNA polymerase:

- TTGACA - 35 bases from the start site of the transcribed gene;
- TATAAT - 10 bases from the start site of the transcribed gene.

(The - sign means 'upstream' of the start site).

These sequences are crucial for efficient gene expression.

consensus sequence

In contrast, eukaryotes appear to use a variety of promoter sequences which may include a consensus sequence of TATA at -30. Hence, when we try to express a eukaryotic gene in a prokaryotic host, the gene may not have a promoter suitable for gene transcription in the host.

Intron sequences

The coding region of most eukaryotic genes contains sequences of DNA which are not represented in an eventual protein product of the gene. The mRNA transcribed from the gene is processed to remove these introns before translation. In eukaryotes the separation of DNA and ribosomes by the nuclear membrane allows time for these

self-splicing

'self-splicing' reactions to occur in the original mRNA transcript. In prokaryotes, there is no nuclear membrane, the DNA and ribosomes are intimately associated and no time for splicing occurs. Therefore when we try to express a eukaryotic gene in a prokaryotic host the eukaryotic introns in mRNA will not be correctly spliced, if they are spliced at all. The eventual protein product of the gene will thus be abnormal.

The production of cDNA clones helps us to get around these problems. Firstly the cDNA that is produced by reverse transcriptase from processed mRNA will contain no introns. Secondly, the cDNA has no promoter sequence and, therefore, can be ligated 'in frame' to a suitable prokaryotic promoter. Using suitable vectors we can also fuse our cDNA to chosen 'termination sequences' which may improve the stability of the gene product. It will be evident that a similar experimental design is followed when we are using vectors designed for expression of genetic information in eukaryotic cells. Figure 5.4 shows how the region around the cloning site of a mammalian vector is designed to enhance transcription.

SAQ 5.3	What are the advantages of making a cDNA bank rather than a genomic DNA bank from a mammalian cell line? Can you think of any disadvantages? Use Figures 5.1, 5.2 and 5.3 for clues.

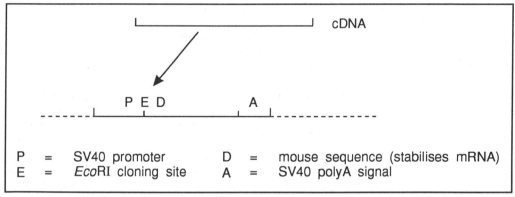

Figure 5.4 Structure around the cloning site of a vector for expressing cDNA in primate cells (pXM). A simian-virus promoter and terminator sequence are used to initiate RNA polymerase binding and finish mRNA synthesis, whilst the mouse sequence produces a stable mRNA. The cDNA is usually ligated to *Eco* RI linkers to allow ligation into the *Eco* RI cloning site.

∏ If the cDNA synthesis can terminate at any time, what is the chance that any one cDNA will be expressed correctly? (Check with Figure 5.2).

The chance will be one in six because there is a one in three chance of fusing the DNA molecules 'in frame' (3 bases/codon) and a one in two chance of the cDNA being the right way round!

5.4 Storage of gene banks

5.4.1 *E. coli* hosts for gene banks

We have already seen how *E. coli* provides the work horse for most recombinant DNA work. When we have ligated our insert DNA into the vector we still may have a tiny amount of DNA for screening and we will very often want to amplify our DNA of interest. *E. coli* is frequently used for this purpose. Even if we are not directly screening for our gene in *E. coli*, we may amplify the recombinant vector/insert DNA in this organism in order to produce enough DNA to manipulate later. This may be especially important when transformation frequencies of an alternative host are poor and we thus need a relatively large amount of recombinant DNA to start with.

However, this process has some pitfalls. *E. coli* wild-type contains a number of enzyme systems that may alter incoming DNA. These are:

- restriction enzymes (DNA may be cut and deleted and so key sequences may be lost);

- modification enzymes (DNA may be methylated at some sites, so altering restriction sites);

- recombination systems (DNA may be 'looped out' and deleted by the recombination enzymes of the *E. coli* host. This is a particular problem where insert DNA contains repeated sequences).

rec⁻, mod⁻, restriction⁻ mutants

Hence, introducing a representative gene bank into such a host could result in the loss of many of our sequences! The approach to this problem relies on the use of particular *E. coli* mutant hosts. Each of the above enzyme systems is coded for by particular genes and so selection of *rec⁻* (recombination) and *mod⁻* (modification) and restriction⁻ (restriction) mutants can provide relatively safe havens for our gene bank.

Do bear in mind though that we are not home and dry yet. Many of the *E. coli* strains most commonly used as gene bank hosts were shown during the late 1980s to have a previously unidentified restriction enzyme system that was degrading some bank DNA.

Additionally, when we grow our *E. coli* host cells containing the gene bank we are applying a strong selection pressure for the fast growing clones at the expense of any clones that adversely affect growth rate.

Finally, the transformation of *E. coli* cells with DNA is always easier with smaller molecules and again we are subconsciously selecting for certain DNA constructs.

In conclusion then, we probably never achieve the ideal of a truly representative gene bank. The work which goes into making a gene bank is enormous and once made the bank is precious! Therefore, it is usual to have separate pools of the bank, if possible, and to preserve them in the fridge or freezer as naked DNA, transformed *E. coli* cells or phage particles where appropriate.

5.5 Screening systems for gene banks

5.5.1 Methods

There are a huge number of screening systems devised for a range of organisms and often these are 'custom designed' for a particular application. It is at this stage that the creativity of the gene cloner really comes to the fore! The question is usually how we can cut down on the workload of screening a very large number of transformants. We shall deal with three of the most widely applicable methods here:

- complementation screening;
- hybridisation screening;
- immunological or protein screening.

5.5.2 Complementation screening

This method relies on our ability to construct mutants of a wide variety of organisms that have some recognisable defect in growth. Such a defect can of course be corrected when we introduce a normal copy of the defective gene into the cell, a process known as complementation. Upon introducing a gene bank, in which such a normal copy of the defective gene is present, into the mutant cells we can select for those clones that can grow under selective conditions. These cells then represent the recombinants in our bank that have a wild-type or normal copy (allele) of the mutant gene.

Let us think about a simple example. Suppose we wished to clone a gene involved in the arginine biosynthetic pathway in *E. coli* (see Figure 5.5).

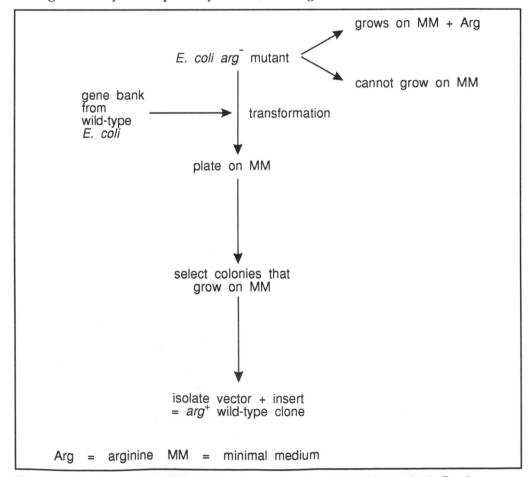

Figure 5.5 A simple strategy for cloning by complementation of an auxotrophic mutation in *E. coli*. Recognition of a wild-type clone depends upon its ability to promote growth on minimal medium that will not support growth of the auxotrophic mutant.

- first isolate an arginine-requiring mutant that cannot grow on normal minimal medium without added arginine. This strain is defective at some stage of the arginine biosynthetic pathway;

- transform this strain with a gene bank made from a wild-type strain that can grow on minimal medium without arginine added.

- plate the cells of the transformed auxotroph onto minimal medium without arginine;

- select colonies that grow on the minimal medium.

complement-
ation

Thus, in this case, we have recognised and selected our clone on the basis of the cell gaining a specific phenotype and so we rely on efficient gene expression. The cells which show wild-type growth on minimal medium now have a normal copy of the gene to correct the defective version. This is known as complementation. The vector and insert can now be isolated from the wild-type clone host and characterised further, perhaps by restriction mapping or sequencing.

| SAQ 5.4 |

Will all the cells that grow on MM be transformed clones? Explain your reasoning.

5.5.3 Hybridisation screening

All hybridisation screening techniques rely on our ability to construct a gene probe which is capable of hybridising with a specific recombinant molecule in our gene bank. The positive clone is identified using a radiolabelled or, less frequently, a non-radioactive probe. Hybridisation screening is thus based on the ability of single-stranded DNA sequences to bind together (or anneal) if they are homologous, ie produce matching Watson and Crick basepairs (A with T and G with C). It is possible to anneal molecules that are slightly mismatched by altering the hybridisation conditions such as buffer constituents or temperature.

Probe construction

There are a variety of methods available for making probes. All of them rely on our ability to reproduce *in vitro*, the natural replicative process of DNA polymerase using a specific DNA substrate and incorporating a label into the newly synthesised strand. In the primer extension method, the DNA polymerase uses the 3' end of a short primer sequence, which attaches to a specific position on the substrate DNA, to start the polymerisation of the second, labelled DNA strand.

primer
extension
method

The Klenow fragment of DNA polymerase I is the favourite choice of enzyme. This fragment lacks the 5'-3' exonuclease activity which may degrade the primer we need to initiate the reaction. Figure 5.6 illustrates the primer extension method.

First the double-stranded DNA substrate is denatured, usually by boiling. Then the primer is added. The primer is a defined, short-sequence of DNA which is homologous to part of the substrate, single-stranded, and binds to it by base-pairing. The DNA polymerase (Klenow) enzyme uses the primer to initiate polymerisation of the new strand of DNA. The buffer solution contains the nucleoside triphosphate DNA precursors, one of which will carry the radiolabel (usually $dCT^{32}P$). Eventually, the probe can be released by denaturing the construct. The labelled probe can be further purified by separating out the unincorporated label on a Sephadex column.

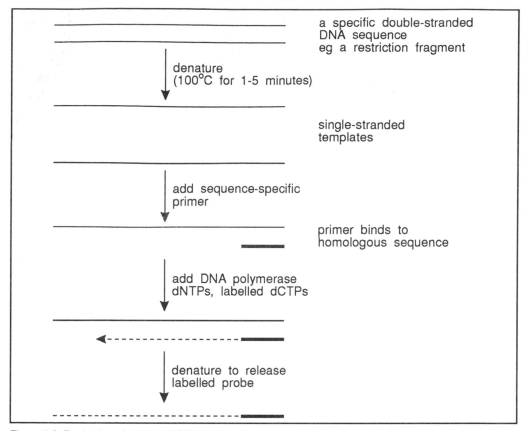

Figure 5.6 Production of a labelled DNA probe using the primer extension method.

Screening filters

The standard hybridisation protocols commonly used all exploit the potential of nylon or nitrocellulose filters to absorb and retain DNA. Ideally, the filters will absorb DNA from plaques - the zones of clearing and lysis on bacterial lawns caused by phage infection. Alternatively, DNA from bacterial colonies can be absorbed, giving a less clear 'blot'. Screening of plaques is easier and clearer than colony blotting where smearing of colonies and the limited numbers per plate can be a problem.

- Replicate Petri dishes are produced which contain transformed cells or plaques which represent our gene bank.

- Nitrocellulose or nylon filters are placed on the plate surfaces for a short time (30 seconds - 1 minute) to absorb the DNA. When blotting cell colonies it may be necessary to disrupt the cells first.

- The filters are removed, denatured with NaOH, washed and finally fixed using heat crosslinking for nitrocellulose filters or UV crosslinking for nylon filters.

- The labelled probe is added in a buffer and the hybridisation mix is incubated for 4-24 hours. The temperature and salt concentration of the incubation buffer both affect how specifically the probe will bind to the target DNA. This is known as the 'stringency'.

• The filters are washed and autoradiographed. Positive clones will be identified as darkened areas of the film caused by the radiation from the radioactivity of the bound probe. The corresponding colonies or plaques can then be identified on the original replicate Petri dishes. Nylon filters are particularly useful in that 'old' probes can be stripped from the filter, leaving the screened DNA behind and ready for re-probing. This also allows for subsequent screening for different clones.

∏ Why are the filters treated with NaOH?

To provide a denatured, single-stranded DNA substrate which allows probe attachment (annealing).

SAQ 5.5

Using the information in Section 5.5.4 put the screening steps into sequence in the diagram shown in Figure 5.7. Explain your reasoning. Note that some steps may be used more than once.

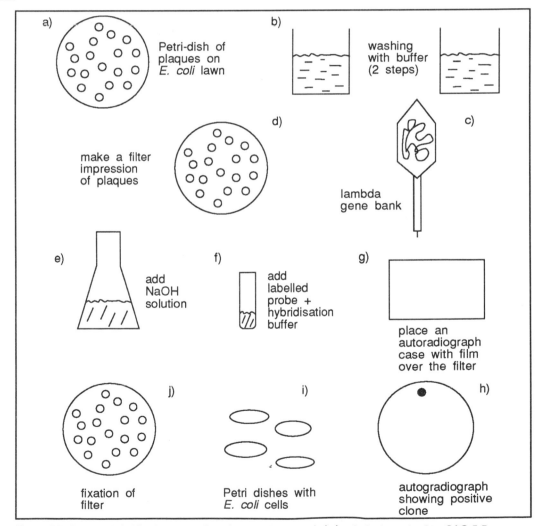

Figure 5.7 Steps in hybridisation screening of a gene bank made in bacteriophage λ. See SAQ 5.5.

5.5.4 Immunological or protein screening

The hybridisation methods described above rely on using a nucleic acid probe (usually DNA and rarely RNA) to detect a specific, homologous DNA sequence in a gene bank.

Screening genomic banks in this way directly examines individual cloned DNA fragments. Screening cDNA banks with hybridisation methods indirectly examines gene products because cDNA is made starting with the mRNA product of transcribed genes. We may also wish to construct a genomic or cDNA bank which allows end clones to be transcribed into mRNA and subsequently translated into protein. In these situations we can use immunological or protein screening methods allowing us to screen clones directly for the final translated protein. Screening relies on the detection of the protein with an antibody. In principle, the screening procedure is similar to hybridisation screening with nucleic acids but protein specific antibodies are used instead of nucleic acid probes.

The outline of the procedure is as follows:

- a protein (the one we are seeking as a gene product from the library), is injected into, for instance, a rabbit. The rabbit's immune system will make antibodies to this protein, blood samples are drawn and the antibodies are purified;

- replicate Petri dishes with recombinant colonies or plaques are transferred to polyvinyl (PVC) or nitrocellulose membranes. The membrane may be pre-coated with the antibody or the antibody may be added after treatment of the cells with chloroform;

- wash the filters to remove unbound probe;

- identify the clones that bind antibody by autoradiography, (if the antibody is radiolabelled) or by using a second antibody directed against the first one. Alternatively, a specific bacterial protein which binds immunoglobulins may be used if the first antibody is not labelled (of course the second antibody or the bacterial protein must be made 'visible' in some way or another).

The chloroform treatment destroys the cell membranes but not the cell walls. Procedures that cause true cell lysis are not used as they will produce artifacts. Usually, bovine serum albumin (BSA) is added to the washing buffer to block non-specific protein binding sites.

5.6 Direct DNA synthesis and isolation

In the foregoing sections we have seen that during cloning and selection procedures we often need short DNA sequences with a specific structure. In this section we will describe how such oligonucleotide sequences can be made.

5.6.1 Oligonucleotide synthesis

oligonucleotide synthesisers

It is now a routine procedure to artificially synthesise short DNA sequences, oligonucleotides or 'oligos', using a series of controlled reactions. Oligonucleotide synthesisers can perform these reactions in a pre-programmed way.

The synthesis reaction relies on the production of a growing chain of nucleotides which are protected until the next nucleotide is added. Figure 5.8 outlines the process and we can discriminate the following steps:

dimethoxy-
trityl

- the initial nucleotide with its 5' end protected, eg with dimethoxytrityl, is bound to a silica support at its 3' end;

dimethyl-phos-
phoramidite

- the second nucleotide is protected at the 5' and 3' ends (3' with dimethyl-phosphoramidite);

- the 3' protectant is removed from the 2nd nucleotide, the 5' protectant is removed from the first nucleotide, and the components are mixed. The time of removal of the protectants has to be carefully controlled;

- the second nucleotide condenses with the first one to form a bi-nucleotide with the first;

- the process is repeated until the oligonucleotide has the desired length and then the chain is cleaved from its support;

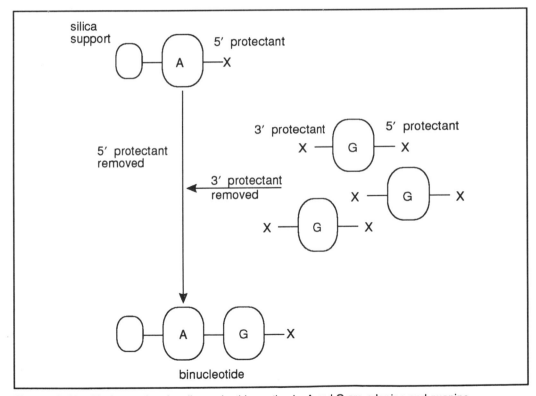

Figure 5.8 Simplified procedure for oligonucleotide synthesis. A and G are adenine and guanine nucleotides respectively.

Π   Why do we need to use the protectant molecules? Look at the diagram in Figure 5.8 and try to find the answer.

The protectant groups are attached to the 3' and 5' ends to prevent reactions between the individual mononucleotides ie in the example to prevent chains of polyA or polyG from forming.

There are two common applications of 'oligos':

- they can bind to specific single-stranded DNA sequences and thus be used as primers to initiate DNA synthesis on a specific template;

- they can be used as probes in screening procedures; for this purpose the oligonucleotide must carry a label.

5.6.2 Polymerase chain reaction (PCR)

We can also synthesise specific oligonucleotide sequences by using the enzyme DNA polymerase. A technique based on this principle is the Polymerase Chain Reaction. With this technique we can specifically and highly accurately amplify a particular DNA sequence without needing a cloning vector. The principle of PCR is the same as that of the primer extension technique shown in Figure 5.6. In theory, the primer extension technique could allow us to amplify a specific probe by repeating the synthesis - denaturation cycle over and over again. However, as the enzyme would denature as well during the denaturation stage, we would have to add fresh enzyme and primer prior to each synthesis stage, which would in practise result in a heavy workload and rapid sample dilution. Moreover, the synthesis reaction might only terminate when it reaches the end of the DNA template, while we may be interested in a shorter DNA sequence.

These disadvantages have been overcome in the PCR technique which uses a heat-stable DNA polymerase from the thermophilic bacterium, *Thermus aquaticus*. The use of a heat-stable enzyme allows repetition of synthesis-denaturation cycles without needing to add more enzyme. Furthermore, the use of two different primer molecules that bind to opposite strands of the DNA and so delimit the sequence to be amplified, allows specific and relatively short sequences to be amplified.

Figure 5.9 outlines the principles of the PCR technique which comprises a series of cycles. Each cycle consists of 3 key steps performed at different temperatures in the presence of nucleotide triphosphates:

- the double-stranded DNA is denatured by heating (eg 95°C for 1 minute);

- the primers are annealed to their complementary target sites (eg 55°C for 1 minute with the optimum temperature being determined by the precise base composition of the DNA sequence);

- the *Taq* polymerase synthesises a complementary DNA chain in a 5' → 3' direction using the primer to initiate the reaction (eg 72°C for 1-3 minutes).

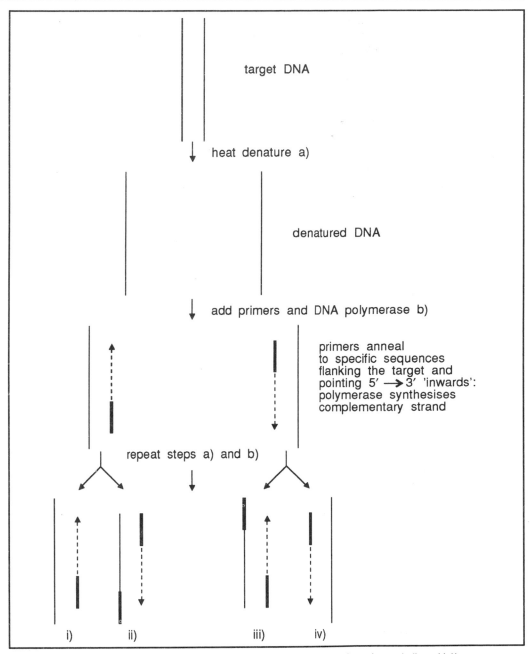

Figure 5.9 The polymerase chain reaction (PCR). Longer primer extensions (strands i) and iv)) are produced on the original template and increase arithmetically (20 copies in 20 cycles). Primer terminated copies (strands ii) and iii)) increase rapidly (10^6 copies of each template in 20 cycles).

5.6.3 Applications of PCR

In general terms, PCR techniques may be applied whenever we wish to produce large numbers of copies of a DNA molecule for which we have sequencing data. The applications are many and varied but we will briefly consider two examples:

- inverse PCR - for amplifying sequences adjacent to a known sequence. This may be of value in isolating clusters of genes of related function which are often grouped together on the bacterial chromosome. We could think, for instance, of the genes responsible for antibiotic production in *Streptomyces spp.* Figure 5.10 illustrates the principle of the procedure. First the DNA molecule of interest is cut with a restriction enzyme at two points some distance removed from both ends of the known sequence. Then, the linear molecule so obtained is ligated to produce a circular molecule and subsequently the circular molecule is cut with a restriction enzyme specific for a site in the known sequence. The resulting linear molecule which now has known sequences at its ends can be amplified using PCR;

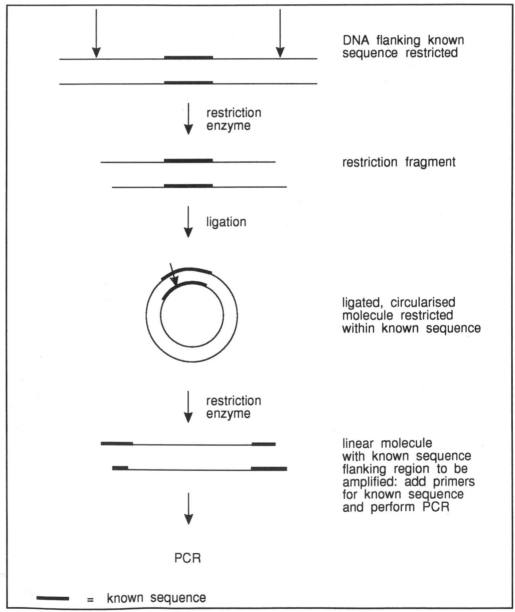

Figure 5.10 Inverse PCR for amplification of a region next to a known sequence.

- amplification of small amounts of DNA for forensic use as PCR can start with tiny traces of DNA - in theory the DNA from a single sperm or blood cell would suffice - and produce sufficient DNA to allow a genetic fingerprint to be made. It may be used to link a particular sample with a specific person.

5.6.4 Timescales for screening methods

As we suggested at the outset, screening can be a lengthy and repetitive procedure. In theory, with some commercial kits it is possible to make a genomic or cDNA bank in less than two weeks. We should, however, have in mind that an enzyme reaction is never 100% efficient and that furthermore, each step has to be successfully completed before the next can proceed. Screening can be a matter of luck - your first experiment may hit the jackpot or you can slave away for months without any success.

PCR stands alone in offering the potential for amplifying specific sequences in less than a day. However, it can take some time to discover the optimum conditions for each system. Remember, we should always run control reactions such as one which is free of target DNA to check for contaminating DNA molecules. The amplified DNA has to be checked to be sure that it is the required sequence.

Finally it is worth remembering that PCR machines are expensive!

SAQ 5.6	List four ways for checking the DNA amplified in a PCR experiment was of the correct sequence.

5.7 Screening; isolation and initial characterisation of clones

First summarise the relative merits of the screening techniques we have met so far by completing SAQ 5.7.

SAQ 5.7

Complete the spaces in the following table that compares the advantages of: PCR, complementation screening, nucleic acid hybridisation and immunological/protein screening.

Method	Advantages	Disadvantages
1) _____	gene function confirmed no probe needed	need to transform cell need suitable mutation need gene expression need to confer selectable phenotype
2) _____	very quick no cloning vector	need sequence data contamination non-specific clones expensive equipment
3) _____	gene function apparent no selectable phenotype needed	probe source? needs gene transcription and translation non-specific clones
4) _____	easy with phage easy to re-probe no selectable phenotype needed gene expression not needed	probe source? tricky with cells non-specific clones function not sure

Table 5.2. A comparison of the strengths and weaknesses of screening methods.

5.7.1 Initial characterisation

After isolation, potential clones must always be rechecked - this is particularly important in the case of hybridisation screening where false positives are common. Once a potential clone is identified by any screening method, we are in a position to begin a full molecular analysis of the gene in question.

When the cloned gene represents most of the insert of a plasmid vector, we may now have a complete sequence and its characterisation would then proceed by direct analysis via restriction mapping followed by sequencing (with around 1-2 kb being convenient for the latter method). Where a smaller gene is carried on a vector with a large insert capacity, like a cosmid, phage or YAC (yeast artificial chromosome), subcloning of restriction fragments of the insert will be required to localise the gene itself.

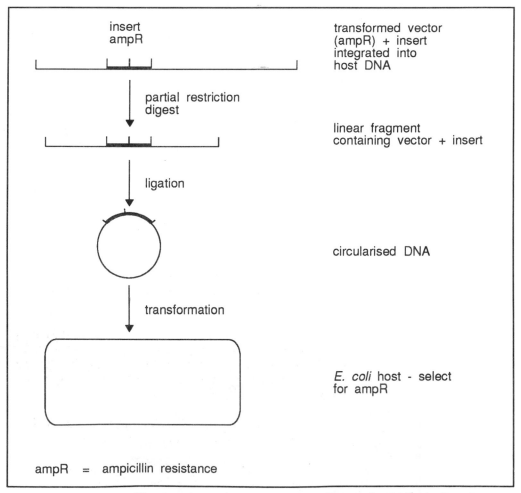

Figure 5.11 Marker (plasmid) rescue: the technique relies on the ability to excise vector plus insert intact. Unless we know that there are no restriction sites within the transformed sequence for the enzyme chosen, many excised molecules will be deleted - others will have large additions of host DNA.

5.7.2 Marker rescue

After complementation screening of prokaryotes, usually we simply lyse the transformed cells and isolate the vector with its insert (see Chapter 4). However, in most eukaryotes, the situation is more complicated. In eukaryotes, transformation is usually integrative which means that the transformed DNA integrates into the host chromosomal DNA. We are then faced with the problem of separating the vector + insert away from the host DNA, a procedure called 'rescue'. Traditionally, the rescue procedure consisted of performing partial restriction digests of the transformant DNA, followed by ligation and hoping to recircularise a molecule that could be transformed and amplified in *E. coli* (Figure 5.11). However, when the transforming molecule is a cosmid it is often possible to rescue the vector with an intact insert, by using a natural process based on the identification of *cos* sites.

integrative transformation

Π   What is the name of this procedure whereby *cos* site specified lengths of DNA are processed into phage particles?

The process is called *in vitro* λ packaging. This process allows the packaging of DNA between *cos* sites spaced by a 38-52 kb DNA sequence. The resulting infective particles can be introduced into *E. coli*. The efficiency of the process is low, perhaps 1-1000 packaged units per μg of total DNA added to the reaction (see also Chapter 4).

After rescue and amplification it is always important to perform a re-transformation experiment in which the rescued, amplified clone is transformed back into the original mutant strain to check for complementation. Complementation screening of eukaryotes can present still other problems. Integration of transforming DNA can lead to many copies per genome and it can also place the clone under the regulatory controls of the host. Such gene dosage and regulatory effects can make clone recognition very difficult.

SAQ 5.8

Choose the most appropriate isolation or screening methods for the following cloning problems briefly explaining your reasons. You may choose more than one method.

1) You wish to isolate a mutant allele of a cloned and sequenced gene which controls antibiotic production in a *Streptomyces spp*. The aim is to find the site of the mutation in the abnormal sequence.

2) It is necessary to clone a gene which controls synthesis of a vitamin in yeast. You hope that by learning more about this gene you may develop the microbe as a production system for the vitamin. You have no idea of the gene's structure but some characterisation of mutants auxotrophic for this vitamin has been carried out.

3) You wish to isolate a fungal, regulatory gene and you know that part of the gene shows strong sequence homology to a part of a mouse gene.

4) You wish to isolate a preproinsulin production gene from mouse. Our knowledge of insulin structure provides us with DNA sequence data and antibodies.

Summary and objectives

In this chapter we have introduced the most common approaches to the isolation of a gene. We have emphasised the importance of constructing a representative gene bank. The differences between genomic and cDNA banks have been explained and their applications outlined. Finally, we have explored several methods for the screening of gene banks for specific clones and have described an alternative method for gene cloning based on PCR.

Now that you have completed this chapter you should be able to:

* explain how genomic and cDNA banks are made;

* understand the criteria for the selection of restriction enzymes;

* calculate the number of clones required to make a representative genomic DNA bank or cDNA bank;

* compare the advantages and disadvantages of constructing genomic DNA and cDNA banks;

* explain how DNA banks can be screened;

* interpret the results of a complementation screening experiment;

* understand the sequence of steps in hybridisation screening of a gene bank;

* understand the procedure of immunological screening;

* explain the PCR method;

* propose methods for confirming the identity of amplified DNA in PCR experiments.

In vitro mutagenesis

In vitro **mutagenesis**

6.1 Introduction

Mutagenesis can be defined as the process of producing structural alterations in DNA (or RNA for some viruses). Very often mutations will give rise to a changed phenotype, but mutations can also be silent, which means that the changed nucleotide sequence of DNA is not reflected in a changed phenotype.

In vitro mutagenesis can be used as a research tool in studying DNA structure and function. Mutagenesis *in vitro* also holds great promise in protein engineering as a method for producing desirable changes in proteins.

The chapter starts with a description of the different types of mutations that are known and then focuses on the methods used to bring about deliberate mutations. First chemically induced mutagenesis is dealt with and after this follows a description of oligonucleotide directed mutagenesis. In this method, relatively short oligonucleotide sequences complementary to specific DNA sequences are used to direct mutagenesis. Then a section on the methods used to introduce short insertions or short deletions is presented and finally random mutagenesis using *Avian myeloblastosis* virus reverse transcriptase is described. The chapter ends with a section on the use of *in vitro* mutagenesis. Note that we will describe the use of a wide variety of enzymes in this chapter. We have provided a list of some of the important enzymes in Appendices 1 and 2. These appendices also summarise some of the properties of these enzymes and their uses in manipulating nucleic acids. Use these appendices as a reference source to remind yourself of their properties.

6.2 Terminology of mutagenesis

Let us first look at the various types of mutation that can occur so that you can learn some of the terminology. We can discriminate five main types of mutation: insertions; deletions; point mutations; translocations and inversions.

- **insertions**: Insertion mutations are caused by the introduction of a piece of DNA (which may even be as small as a single basepair) into the DNA sequence.

- **deletions**: Deletion mutations are caused by the removal of a section of DNA (or even a single basepair) from the DNA sequence.

- **point mutations**: basepairs in DNA may be altered. An alteration that changes only a single basepair is termed a point mutation. Insertion and deletion of single base pairs can also be considered to be point mutations. Other types of point mutation are also possible, the most common class being the transition. In a transition mutation one purine or pyrimidine is replaced with the other purine or pyrimidine respectively. Thus, in a transition an A-T base pair is replaced with a G-C pair. A less common point mutation is the transversion, in which a purine is replaced by a pyrimidine or *vice versa* causing, for example an A-T basepair to become a T-A or C-G basepair.

- **translocations**: these are mutations in which a nucleotide sequence is moved from a specific location within the DNA into another location. The translocated segment may be short but often consists of quite a long nucleotide sequence. The movement of a segment of the DNA can give rise to marked changes in phenotype.

- **inversions**. Inversions are mutations in which a nucleotide sequence is removed from the DNA and re-inserted in the opposite orientation. We can represent this situation in the following way:

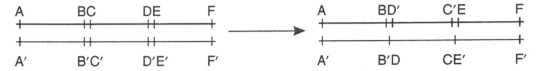

| **SAQ 6.1** | Indicate whether each of the following mutations are deletions, insertions, inversions, transitions or transversions. |

1) A G-C basepair is converted to an A-T basepair.

2) An A-T basepair is converted to a G-C basepair.

3) A G-C basepair is converted to a T-A basepair.

4) The sequence GCTCGT is converted to GCTCT.

5) The sequence GCTCGT is converted to GCTCGGT.

6) The sequence GCTCGT is converted to GCTTGTC.

7) The sequence G C T C C G G T C is converted to G C C C G G A T C .
             C G A G G C C A G   C G G G C C T A G

All of the types of mutations described above occur *in vivo* and can often be produced *in vitro*. During *in vitro* mutagenesis, specific mutations can be introduced at specific sites in the target DNA sequence. A variety of techniques is available for bringing about mutations *in vitro*. These techniques include treatment with chemicals (which usually produces point mutations) and the use of genetic engineering (which can be used to produce point mutations and larger alterations in DNA sequences).

We will first deal briefly with the use of chemicals to produce mutations and then describe the rapidly expanding fields of oligonucleotide directed mutagenesis and exonuclease mutagenesis.

6.3 Chemical mutagenesis

Many chemicals are available that will mutagenise both double- and single-stranded DNA. These include nitrous acid (which converts G-C basepairs to A-T basepairs and vice versa), hydroxylamine (which only converts G-C basepairs to A-T basepairs), hydrazine (which causes all possible basepair changes), acridines (which cause the addition or deletion of from one to at least twenty bases) and sodium bisulphite which leads to the formation of uracil in DNA by deamination of cytosine. Replacement of cytosine by uracil will result in incorporation of a complementary A residue during DNA synthesis, which eventually leads to the original G-C basepair being replaced with an A-T basepair.

The use of chemicals to produce mutations is generally declining due to the low frequency at which mutants are recovered and the generally limited range of mutations that can be produced. Sodium bisulphite does though still enjoy some success as a mutagen. Sodium bisulphite is usually only considered to act on single-stranded DNA. Single-stranded DNA (we will explain later in this chapter how single-stranded DNA can be produced) is treated with a slightly acidic solution of sodium bisulphite. As we have already seen this causes the deamination of cytosine residues (forming uracil) and the production of transition mutations after DNA replication. Special bacteria, that do not destroy DNA containing uracil, are used as hosts for the replicating DNA.

mutagenesis by sodium bisulphite

Sodium bisulphite mutagenises DNA quite efficiently, but can only produce transition mutations, which often result in conservative amino acid substitutions. The use of sodium bisulphite is therefore quite limited. Sodium bisulphite can also produce mutations in double-stranded regions of DNA outside the target site and the use of techniques to ensure that the mutation is in the expected region is therefore compulsory. This does of course lead to a more complicated and time consuming process.

SAQ 6.2

Listed below are some mutations similar to those in SAQ 6.1. For each mutation given, name the type of mutation and the chemical mutagen that could produce it.

1) An A-T basepair is converted to a G-C basepair.

2) A G-C basepair is converted to an A-T basepair.

3) A G-C basepair is converted to a T-A basepair.

4) A G-C basepair is converted to an A-T basepair via mutagenesis of single-stranded DNA.

5) The sequence at ATCTGA is converted to ATCGTGA.

6.4 Oligonucleotide directed mutagenesis

6.4.1 General strategy

The words 'oligonucleotide directed mutagenesis' cover a wide range of powerful techniques used to add, delete or substitute basepairs in a piece of DNA of known sequence. The mutation can be directed against any particular basepair or region of the target sequence and is thus a type of site-directed mutagenesis.

These techniques depend, as the name would suggest, on the use of chemically synthesised, short sections of DNA (oligonucleotides) to introduce the desired mutations. Perhaps the easiest way to visualise how these techniques work in principle is to study the scheme shown digrammatically in Figure 6.1. The steps are described below:

• clone the DNA sequence to be mutated into a vector capable of producing single-stranded DNA (such as bacteriophage M13 vectors);

- prepare single-stranded DNA from the recombinant M13 vector. This strand is referred to as the +strand;

- design and synthesise oligonucleotides carrying the desired mutations. (We have labelled this on our diagram as a synthetic oligonucleotide carrying a mismatch nucleotide; this oligonucleotide will hybridise with the target sequence except at the mutated point);

- hybridise the mutagenic oligonucleotide to the target DNA;

- extend the hybridised mutant oligonucleotide with DNA polymerase;

- transfect suitable host bacteria such as *E. coli*;

- screen the plaques produced by hybridisation using the mutagenic oligonucleotide to select the phages carrying the mutation;

- prepare single-stranded DNA from the mutagenised phage and sequence to confirm that it carries the desired mutation.

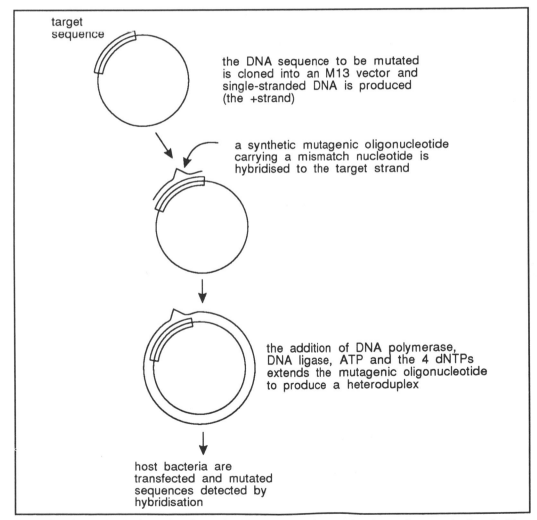

Figure 6.1 Generalised scheme for oligonucleotide directed mutagenesis using a phage vector (see text for details).

Many of these techniques will have been described elsewhere and are common procedures in molecular biology. There are though special considerations when applying these techniques to oligonucleotide direct mutagenesis. We will look at these in more detail.

Target DNA preparation

size of target
DNA is
important

The size of the target DNA in the M13 or equivalent vector should be kept to a minimum for various reasons. Large DNA fragments tend to be unstable in single-stranded bacteriophage vectors and can undergo spontaneous deletions. Large DNA fragments also increase the chance that the mutagenic oligonucleotide will not be specific for one site in the target DNA. Finally, sequencing of the entire mutated target DNA needs to be carried out in order to confirm that the mutation was as expected. Obviously the smaller the section of DNA that needs to be sequenced, the quicker it is achieved.

Oligonucleotides

Obviously the mutagenic oligonucleotide must be capable of hybridising to the target sequence in the cloned DNA sequence. To achieve this some potentially conflicting interests have to be reconciled.

palindromes

Generally the mutagenic oligonucleotide should be free from features that lead to the formation of secondary structures, such as palindromic or self-complementary sequences. Such structures reduce - sometimes most dramatically - the efficiency of mutagenesis.

specificity of
the mutagenic
oligonucleotide

The mutagenic oligonucleotide should also be designed so that it is not capable of producing stable hybrids with the vector DNA or regions in the target DNA other than the sequence used to design the oligonucleotide.

double-primer
method

If the double-primer method is to be used, a second standard primer is added as well as the mutagenic primer to make the process quicker (see Figure 6.4). In this case, the 5' terminus of the mutagenic oligonucleotide should be able to form a perfect, stable hybrid with the template so that DNA synthesis initiated from the upstream standard primer does not displace the mutagenic oligonucleotide. A number of 8-10 perfectly matched nucleotides are required at the 5' terminus if the Klenow fragment is used. An

use of T4
polymerase or
a 'Sequenase'

alternative strategy can be adopted by the use of T4 polymerase or a 'Sequenase' that does not readily displace the mutagenic oligonucleotide. The 3' terminus of the mutagenic oligonucleotide should be capable of producing hybrids with the template DNA that are sufficiently stable so as to allow priming of DNA synthesis at high efficiency. If no stable hybrid is formed the 3' terminus of the oligonucleotide will be susceptible to exonucleolytic degradation by the Klenow fragment of DNA polymerase I. Therefore about 7-9 perfectly matched nucleotides are required at the 3' terminus.

The difference in thermal stability between the hybrids formed by the mutagenic oligonucleotide and wild-type DNA and that formed by the mutagenic oligonucleotide and mutated target DNA should also be sufficient to allow screening with the mutant oligonucleotide to identify potential mutants. As oligonucleotides increase in length the difference in thermal stabilities between perfectly matched hybrids and ones containing a mismatch decreases.

∏ Before reading on see if you can mention two factors which determine the length of the oligonucleotides used for oligonucleotide-directed mutagenesis.

typical sizes of mutagenic oligonucleotides

Mutagenic oligonucleotides have to be long enough to allow stable hybrids to be formed at the 3' and 5' ends, but short enough to allow screening with the mutagenic oligonucleotide. Typically then most oligonucleotides for substitution, addition or deletion of single nucleotides are 17-19 nucleotides long and carry the mismatched base at the centre or immediately 3' of the centre.

If two or more contiguous nucleotides are to be mutated, longer oligonucleotides will be used (usually longer than 24 nucleotides each side of the central mutagenic region).

Mutagenic oligonucleotide hybridisation

temperatures used to form hybrids

Hybrids will form as a mixture containing the mutagenic oligonucleotides and target DNA is allowed to cool, subsequent to heating the mixture in order to remove secondary structures. The temperature at which hybrids form depends on the base composition of the mutant oligonucleotides. Room temperature is usually sufficient, but if the mutagenic oligonucleotide is A-T rich then lower temperatures may be required (12-17°C) in order to form stable hybrids.

Primer extension

When stable hybrids have been formed, a mixture of DNA polymerase, DNA ligase, ATP and the four dNTPs are added and DNA synthesis is initiated from the 3' end of the oligonucleotide(s). The Klenow fragment of DNA polymerase I remains the enzyme that is most widely used, although both T4 DNA polymerase and 'Sequenase' are reported to be more efficient.

Transfection of *E. coli*

The resulting heteroduplex (= DNA duplex in which the two strands of the duplex are not the same) from the hybridisation and elongation steps can be transfected into an *E. coli* host.

Identification of mutants

Plaques produced in lawns of the host bacteria can be probed with the mutagenic oligonucleotide to identify bacteriophages with mutated copies of the target DNA. Usually only a small proportion of bacteriophages carrying the mutated target DNA are found, most carry the wild-type DNA.

The putative mutations can now be sequenced and the mutated DNA subcloned if necessary. You should now be familiar with the basic strategy and methodology of oligonucleotide directed mutagenesis so we will now go on to consider some methods in more detail.

6.4.2 The single primer method

This technique is the simplest one used for site directed mutagenesis. It does though illustrate many of the points that we have covered earlier in this section. The method relies on initiating DNA synthesis from a single oligonucleotide which carries a base mismatch within the complementary sequence. The basic method is shown in Figure 6.2. This figure shows that the synthetic oligonucleotide is incorporated into the resulting heteroduplex molecule and that finally wild-types and mutants will result.

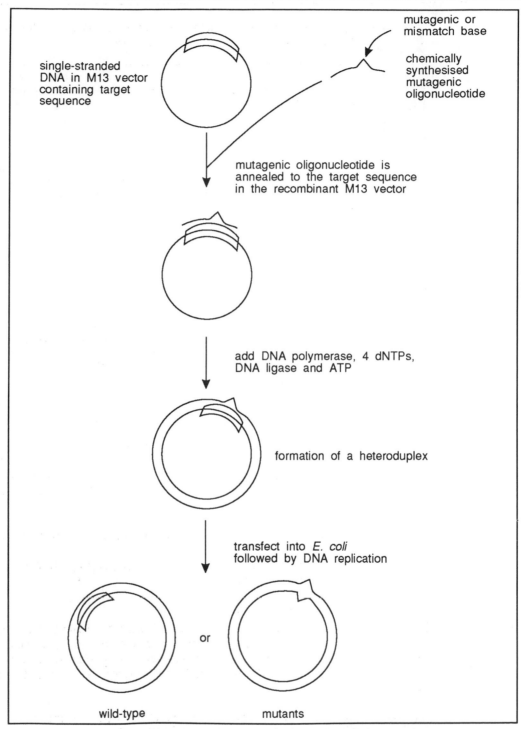

Figure 6.2 Oligonucleotide - directed mutagenesis using a single primer (see text for details).

This single primer method can also be extended so that both deletions and insertions in the target DNA can be made. Figure 6.3 illustrates schematically how this can be achieved. In principle, such a modification can also be used in other procedures such as, for instance, the double primer method (see below).

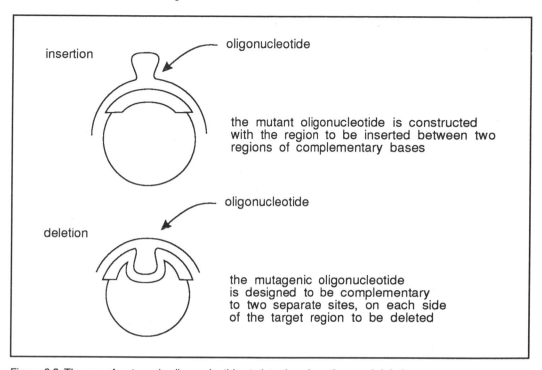

Figure 6.3 The use of mutagenic oligonucleotides to introduce insertions and deletions.

limitations of the single primer method

The single primer method is the simplest method, and, as might be expected, it has several drawbacks, amongst which the contamination of the heteroduplex DNA with template DNA and partially double-stranded molecules is an important one. There is therefore a need to purify covalently closed circular DNA before the host bacteria is transfected.

The single oligonucleotide method is also susceptible to 5'-3' exonucleases.

More advanced techniques have been developed in order to overcome the shortcomings of the single primer method.

6.4.3 The double primer method

double primer method

In the double primer method, DNA synthesis is initiated from two primers, one being the mutagenic oligonucleotide and the other a standard sequencing primer (see Figure 6.4). This results in a much more efficient and quicker procedure as there is no longer a need to isolate covalently closed circular DNA before transfection of the host bacteria.

Of course in the design of the mutagenic oligonucleotide, all the points mentioned in Section 6.4.1 have to be taken into consideration to ensure that stable hybrids are formed and DNA synthesis is uninhibited. Furthermore, the mutagenic oligonucleotide used must be phosphorylated by T4 polynucleotide kinase for reasons that will become apparent shortly.

As with many techniques in which oligonucleotides are used, the synthetic oligonucleotide may need to be purified before use by Sep-Pak C18 cartridges or polyacrylamide gel electrophoresis.

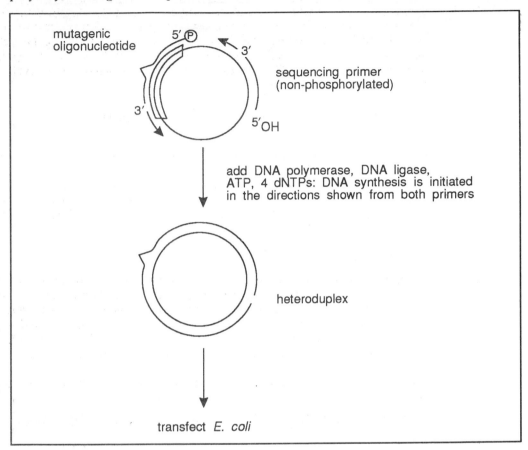

Figure 6.4 The double primer method for oligonucleotide-directed mutagenesis.

The mutagenic primer (which has been phosphorylated) and the non-phosphorylated sequencing primer have to form hybrids with the single-stranded template DNA, this process is called annealing. To produce these, an excess (10-50 fold molar excess) of each primer is used (for example if 0.5 pmol of target DNA is used 10 pmol of each oligonucleotide can be used). The total volume in which the annealing reaction takes place is kept low. The mixture of oligonucleotides and template DNA is then heated to about 20°C above the melting temperature (Tm) of the expected hybrids. The Tm (in °C) of a perfect hybrid between the oligonucleotides and the target DNA depends upon the base composition and the nature (pH, ionic strength) of the solvent in which the DNA is dissolved. The mixture is then allowed to cool and maintained at room temperature (approximately 23°C) for about 5 minutes.

importance of knowing Tm

Then DNA synthesis is initiated from the two oligonucleotides. To do this a mixture of dNTPs, ATP, T4 ligase and the Klenow fragment of DNA polymerase I is added to the annealing reaction mixture. The contents are placed at about 15°C for between 4-12 hours (although the reaction conditions will need to be altered if other DNA polymerases are used).

The host bacteria can now be transfected with the mutated heteroduplex and the plaques in the bacterial lawns can be screened to isolate potential mutated sequences.

disadvantages of the double primer method

Although the double primer method is a big improvement on the single primer method, it still has some drawbacks. As in the single primer method heteroduplices will, upon replication, yield both mutant and non-mutant progeny. A problem with both the single and the double primer method is that heteroduplices may also be subject to DNA mismatch repair by *E. coli,* because *E. coli* has a DNA mismatch repair system that corrects mistakes after DNA replication. This mismatch repair system is directed against under-methylated DNA.

∏   What will be the consequence of this repair system?

importance of DNA repair

The mismatch repair system will be directed against the newly, *in vitro* synthesised strand. The mismatched base(s) in the newly synthesised strand will thus be repaired. It has been demonstrated that certain types of nucleotide mismatch are subject to mismatch repair both *in vivo* and *in vitro.* These factors contribute to the low efficiency of the single and double primer methods, which may be as low as 5% using a single primer.

6.4.4 Gapped-duplex mutagenesis

We have seen in the previous section that important drawbacks of both the single and double primer methods are the production of both mutant and non-mutant progeny and that the *E. coli* repair system is directed against the mutated strand produced *in vitro.* Both drawbacks can be overcome by the use of the so called gapped-duplex method, one of several strand selection methods because the growth of progeny phage containing mutated sequences, as opposed to wild-type sequences, is favoured.

We will now look at several strand selection methods, starting with the gapped-duplex method.

The technique is illustrated in Figure 6.5 and we can follow this through in stages. These stages are:

*   the DNA fragment to be mutated is cloned into an M13 or similar vector;

*   single-stranded DNA that contains the target sequence is isolated;

*   double-stranded DNA of the M13 vector alone is also isolated;

*   the double-stranded M13 DNA is cleaved at the restriction site that was used for cloning the fragments to be mutated into the vector. Then it is denatured and allowed to form a partial DNA duplex with excess single-stranded target DNA. The molecule so produced is referred to as the gapped-duplex;

*   the mutagenic oligonucleotide primer is hybridised to the single-stranded gap, which is of course the target sequence, and the gap is further filled in with the Klenow fragment and DNA ligase;

*   the host bacteria is transfected.

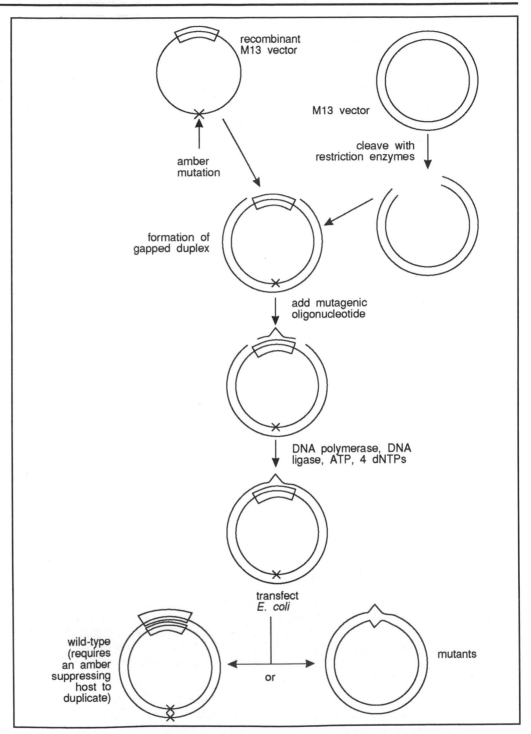

Figure 6.5 Gapped-duplex mutagenesis (illustrating the use of amber mutations). Amber mutants are a class of mutations that result in the creation of a UAG codon in mRNA. This codon normally signifies translation termination, so polypeptide synthesis stops at the amber sites. Such mutations can be suppressed in certain strains of *E. coli* possessing a tRNA with the AUC anticodon. These insert an amino acid at the UAG site and thereby permit continued translation (see text).

⊓ Write down the advantages of gapped-duplex mutagenesis.

This method already exhibits several advantages. The synthetic mutagenic oligonucleotide is unlikely to anneal to sequences within the recombinant M13 genome other that the target sequence since only a small 'window' of single-stranded DNA is available to the mutagenic oligonucleotide for hybridisation. Furthermore since the entire vector sequence is covered by the complementary strand the 'filling in' of the gap is much more efficient.

This is not to say though that the method cannot be improved even further, and we will now describe several further improvements of the method.

amber
mutations

The first improvement we will look at is the use of 'amber' (nonsense) mutations (see legend to Figure 6.5) in the vector which therefore can only be grown in host bacteria that suppress amber mutations. When a strain of host bacteria that cannot suppress amber mutations is transfected only mutant progeny should be recovered. This has resulted in efficiencies of over 70% being claimed.

An advantage of the use of amber mutations is that a selection in favour of the mutant sequence can be applied, hence the phrase 'strand selection'.

⊓ Write down a disadvantage of the use of amber mutation in gapped-duplex mutagenesis.

The use of amber mutants means that if further mutations are needed the mutant sequence has to be recloned into a vector carrying an amber mutation. This is one of the main disadvantages of using amber mutatants.

dam⁻ E. coli
host

There is another advantage that can be incorporated into the gapped-duplex method. This is the use of *E. coli* strains deficient in dam methylase activity (*dam⁻* strains). In these *E. coli* strains, the DNA will be under methylated. The target DNA that is to be mutated is cloned into a M13 vector and grown in a *dam⁻* host strain. This results in the production of unmethylated single-stranded DNA containing target sequence.

⊓ What will be the consequence of the vector containing the target sequence being produced in a *dam⁻* host?

During gapped-duplex formation a heteroduplex is formed that is methylated only on the incomplete strand formed upon restriction of the double-stranded DNA, thus a hemi-methylated duplex is formed. The mutagenic oligonucleotide will become part of the methylated strand. If DNA mismatch repair takes place it will now be directed not against the strand containing the mutagenic oligonucleotide but against the wild-type strand which is unmethylated.

Advances such as these have improved the efficiency of mutagenesis *in vitro* to such a point that it is unnecessary to screen for mutant progeny by hybridisation. Instead a few clones selected at random can be sequenced to select for mutated sequences.

6.4.5 Coupled priming

The coupled priming method is based on the use of a pair of reciprocating selectable markers which enable a second selectable mutation to be introduced as the first is removed. Therefore repeated rounds of mutagenesis can be carried out without the need of re-cloning. The principle of the method is illustrated in Figure 6.7.

Eco B and *Eco* K recognition sites

The markers used are the recognition sites for the *Eco* B and *Eco* K restriction endonucleases. As can be seen in Figure 6.6 the recognition sequences for the two restriction endonucleases are very similar and therefore can be interconverted by site-directed mutagenesis. Furthermore the use of *E. coli* K or B strains allows for specific selection. In an *E. coli* K strain double-stranded DNA with an *Eco* K site will be modified or restricted. Similarly in *E. coli* B strain double-stranded DNA with an *Eco* B site will be modified or restricted. However, hybrid sites that contain an *Eco* K or *Eco* B recognition site in one strand and a mutationally altered site in the other are not recognised by the corresponding restriction or modification enzymes. Together the possibility of selection marker interconversion and strand selection make possible a process of cyclic selection which allows many rounds of mutagenesis.

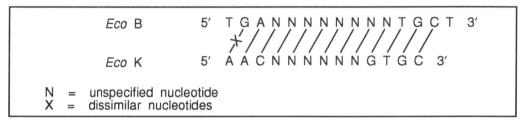

Figure 6.6 Recognition sequences of *Eco* B and *Eco* K restriction endonucleases and modification systems.

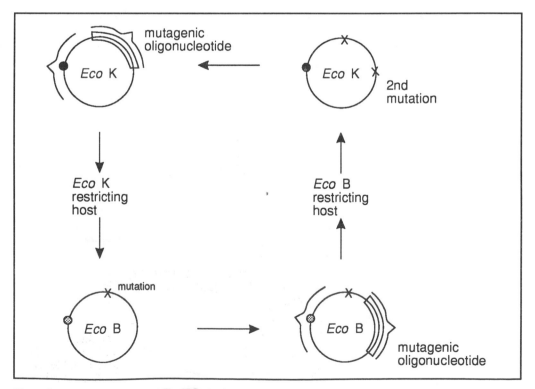

Figure 6.7 Coupled priming and cyclic selection (see text for details).

In the first round of mutagenesis one primer is used to convert the *Eco* K site to an *Eco* B site and one primer is used to introduce the desired mutation. After transfection into a repair deficient K strain host only phages in which the *Eco* K site has been converted to an *Eco* B site will be viable. The formation of the *Eco* B site provides the basis for selection in the next round of mutagenesis, during which the *Eco* B site is converted back to an *Eco* K site and a second mutation is brought about. Following transfection into a B strain host only phages with an *Eco* K site will be viable. Then a third round of mutageneis can be started .

6.4.6 The Kunkel method

Kunkel has developed a method of strand selection which depends on the use of a specialised host which is deficient in two enzymes for the production of single-stranded DNA prior to mutagenesis. The two enzymes in which the host is deficient are dUTPase (*dut*) and uracil glycosylase (*ung*) and the overall effect of these deficiencies is to increase the intracellular pool of dUPT which leads to some dUTP being incorporated into DNA instead of dTTP. Uracil residues are usually removed from DNA by uracil glycosylase, but in *ung*⁻ strains the uracil is not removed and so a number of thymine residues are replaced by uracil residues. If the M13 phage is grown in *dut*⁻ *ung*⁻ hosts then about 20-30 uracil residues are incorporated per genome. If the bacteriophages are then used to transfect an *ung*⁺ strain of bacteria the uracil is removed rapidly, resulting in the production of pyrimidinic sites. This removal of uracil results in the production of lethal lesions, presumably because DNA synthesis is blocked at these sites or because they are sites for special classes of endonucleases.

dut⁻ and
ung⁻ hosts

This then is the basis for a method of site directed mutagenesis which relies on the strong selection against DNA containing uracil (see Figure 6.8).

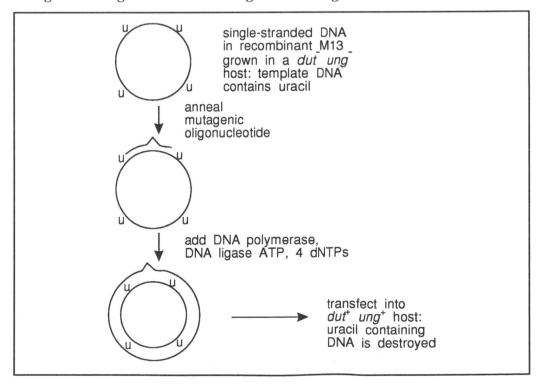

Figure 6.8 The Kunkel method of oligonucleotide directed mutagenesis.

Template DNA is produced by the growth of recombinant bacteriophage M13 in a host that is *dut⁻ ung⁻*. The single-stranded template DNA therefore contains uracil (which does not interfere with manipulations *in vitro*) and is used in a standard mutagenesis procedure to generate heteroduplex molecules with uracil in the template strand and thymine in the strand synthesised *in vitro*.

∏   What will be the result if such heteroduplex molecules are transfected into an *ung⁺* host strain?

Subsequent growth in an *ung⁺* strain results in the destruction of the template strand and the production of phages with only the mutated sequence. Mutation efficiencies of greater than 50% have been reported using this method.

6.4.7 Use of thionucleotides

use of thionucleotides

The methods we have just looked at depend on strand selection *in vivo*. Strand selection can, however, be made *in vitro* and this has several advantages. The method used relies on the fact that certain restriction endonucleases cannot cleave DNA that contains thionucleotides (these are nucleotides where an oxygen on the alpha phosphate is replaced with a sulphur, see Figure 6.9). If DNA synthesis is initiated from the mutagenic primer oligonucleotide in the presence of thionucleotides, a heteroduplex is formed in which only the mutant strand contains thionucleotide residues. The heteroduplex can then be treated with restriction enzymes that will not cleave phosphorothiated DNA. This results in the eventual destruction of the template strand due to exonuclease digestion.

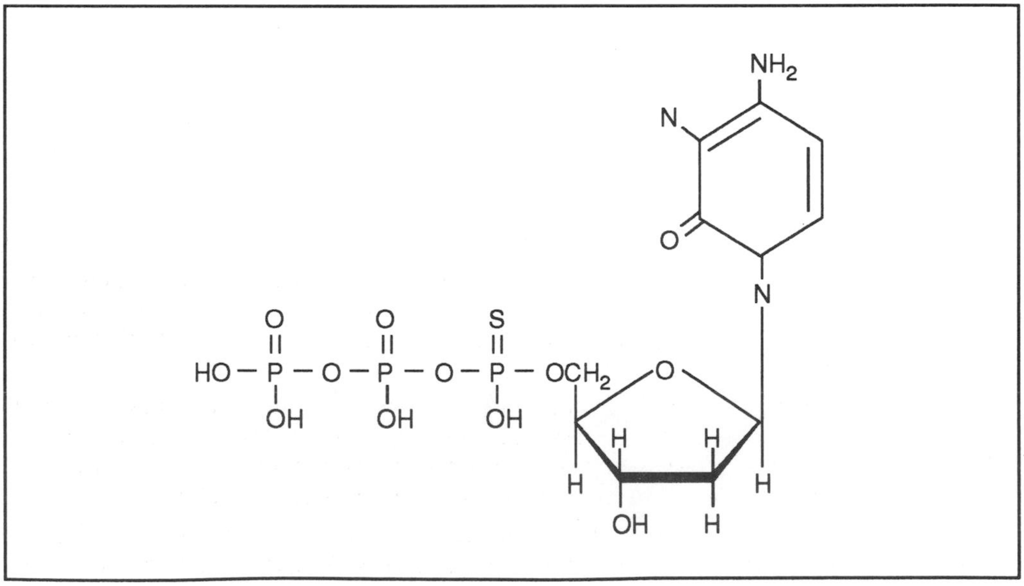

Figure 6.9 The structure of dCTPαS, a typical thionucleotide.

| **SAQ 6.3** | Below are two columns of terms and definitions. Relate the items from the left-hand column to the allied items in the right-hand column. |

Kunkel method

allows multiple rounds of
mutagenesis without re-cloning

allows strand selection *in vivo*

allows strand selection *in vitro*

an advantage of the gapped-
duplex method

dam⁻ strains

vectors capable of producing
single-strand DNA

an enzyme used in producing
a mutant heteroduplex

an enzyme involved in excision
of uracil from DNA

uracil glycosylase

amber mutations

only a small window of single-
stranded DNA is available for
the oligonucleotide to anneal to

cyclic selection

thionucleotides

M13

produce undermethylated
template DNA

DNA polymerase

relies on destruction of DNA
containing uracil

6.5 Methods to introduce short insertions or to produce short deletions

mutagenesis

In the previous sections we described how deliberate changes in the structure of DNA can be introduced using chemical mutagenesis and oligonucleotide directed mutagenesis.

In this section we will look at other ways of producing small deletions or insertions in the target DNA.

6.5.1 Gap-sealing mutagenesis

nick formation

Small sized deletions can be made in DNA by treating double-stranded DNA with agents that introduce a cleavage, usually called a nick, in one strand of the DNA. Agents used for this purpose are pancreatic DNase I and restriction endonucleases in the presence of ethidium bromide.

nick extension

The nick formed by these enzymes can be extended by nucleases such as exonuclease III. The remaining single-stranded region can then be removed by the action of S1 nuclease, after which the DNA can be recircularised. The resulting DNA molecule is thus shortened by the number of nucleotides removed by the exonuclease III digestion.

deletions at
random
positions

As we will see later, the action of DNase I cannot be controlled so that the nicks introduced into the single strand of the target DNA can be at any position. Therefore, the deletions made by this method occur at random positions.

Gap-sealing mutagenesis (illustrated in Figure 6.10) is one of the simplest methods used to generate deletions in pieces of DNA when there are no convenient restriction sites available. The figure also illustrates, some of the basic steps used in other, more complicated methods.

a) using restriction endonuclease in the presence of ethidium bromide
    and exonuclease III

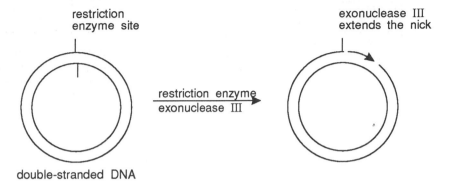

b) using DNase I in the presence of ethidium bromide and exonuclease III

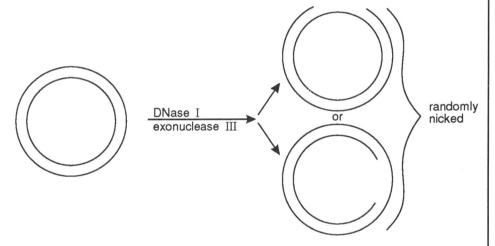

c) the partially single-stranded DNA molecule from a) or b) is treated with
    S1 nuclease and recircularised

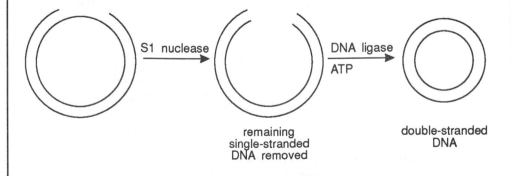

Figure 6.10 Gap-sealing mutagenesis. A nick is made by a restriction enzyme (a) or DNase I (b) and
extended by exonuclease III. Subsequently the single-stranded DNA is removed by S1 nuclease and the
DNA is recircularised with DNA ligase (c).

6.5.2 Linker-insertion mutagenesis

Linker-insertion mutagenesis is a relatively simple but effective way of producing insertions into cloned DNA fragments. The basic principles of the method are outlined in Figure 6.11.

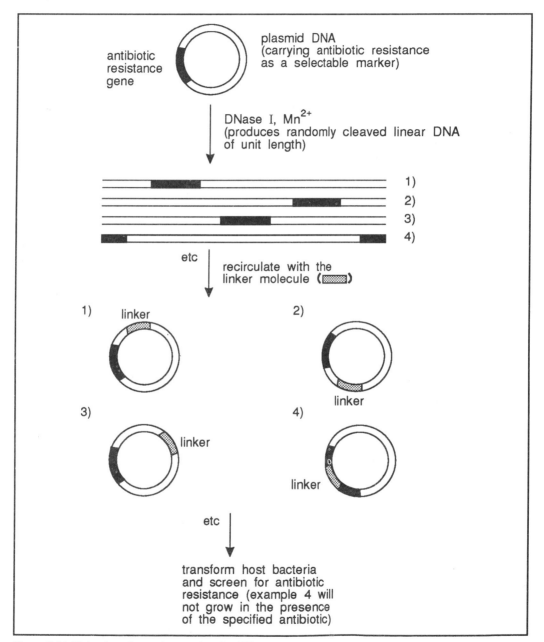

Figure 6.11 Linker insertion mutagenesis using DNase I.

The method relies on the double-stranded cleavage of plasmid DNA that contains the target sequence. After cleavage, a small piece of synthetically produced DNA (the linker) can be inserted to produce a mutation, or to introduce a restriction enzyme site.

The plasmids produced after the DNA is recircularised can be used to transform suitable host bacteria. The presence of an antibiotic resistance gene on the plasmids allows selection of transformed hosts.

The double-stranded cleavage can be produced by:

- the use of DNase I;

- the use of restriction enzymes.

random
cleavage with
DNase I

If DNase I is used, the plasmid DNA is digested with limiting amounts of pancreatic DNase I in the presence of divalent cations such as manganese. Using these conditions the target DNA molecule is cut randomly and unit length linear DNA molecules are produced. These molecules can then be purified to remove undigested plasmid. Subsequently the linearised DNA is recircularised in the presence of the synthetic linker and the recircularised plasmid is used to transform host bacteria. The design of the linker molecule usually is such that it contains a restriction site so that the approximate position of the linker DNA within the target DNA is easily determined by restriction mapping of the plasmid.

If the DNA is linearised by partial digestion with restriction enzymes a similar overall scheme is adopted. The restriction enzyme chosen to linearise the target DNA molecule should be one that cuts DNA frequently and produces blunt ends which allows a wide variety of linkers to be ligated into the target DNA. Enzymes commonly used include *Alu* I, *Hae* III and *Rsa* I.

Both systems have their disadvantages, and for some applications can be replaced by more efficient methods, but for target DNA sequences with a length between about 300 and 2000 basepairs containing only a few restriction sites, this may prove to be the only applicable method.

specific site
deletions

The method can also be adapted to introduce deletions at specific sites in DNA. If the linker contains a restriction site it can be cut with the appropriate restriction enzyme to produce a linear DNA molecule. The nucleotides of the linker sequence and a few from the flanking target DNA sequence can subsequently be removed by the use of exonucleases (alone or in combination, such as exonuclease III, *Bal* 31, S1 nuclease or mung bean nuclease). The resulting shortened DNA molecule can then be recircularised. To produce larger deletions we usually use so-called linker-scanning mutagenesis (see below).

6.5.3 Linker-scanning mutagenesis

Linker-scanning mutagenesis is a particularly elegant, though time-consuming method used to identify DNA regions with a regulatory function. Mutations in such a regulatory DNA region will result in loss of the regulatory function. Linker-scanning mutagenesis is more precise than the production of nested deletions, that can be used for the same purpose. The method basically consists of producing 5' and 3' deletions and ligating these with a synthetic linker which exactly replaces the number of nucleotides removed from the target DNA sequence.

The technique is illustrated in more detail in Figure 6.12 and is described in the following paragraph.

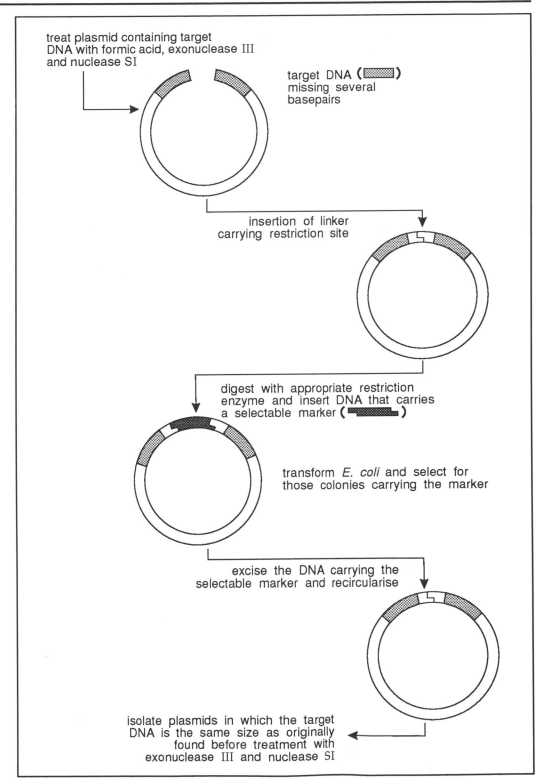

Figure 6.12 Linker scanning mutagenesis.

depurination
with formic acid

Plasmid vector carrying the target DNA is treated with formic acid. The treatment with formic acid results in depurination of the DNA. Depurination occurs more-or-less at random, but the experimental conditions chosen are such that, on average, one base in the target DNA is depurinated. The partially depurinated DNA is then treated with exonuclease III which recognises depurinated sites. Exonuclease III treatment results in the formation of single-stranded gaps. Then nuclease S1 is used to remove the remaining region of single-stranded DNA. After the termini have been repaired, a synthetic linker molecule, containing a restriction enzyme site, is inserted. The resulting DNA is subsequently cleaved at this restriction enzyme site within the linker and a piece of DNA that carries a selectable marker is inserted. This means that upon transfection, plasmids carrying the selectable marker, and thus the linker, can be obtained. After production of the DNA carrying the selectable marker which has now done its job, the marker can be removed and the synthetic linker re-ligated to recircularise the DNA. The modified target DNA thus has approximately the same size as the target DNA before insertion of the synthetic linker, and can be isolated by fractionation on a polyacrylamide gel. The position of the synthetic linker can then be determined by digestion with appropriate restriction enzymes.

insertion of
linker with a
restriction
enzyme site

insertion and
removal of
selectable
marker

As the initial point of depurination is more-or-less random, the linker sequences can be introduced at many points in the target DNA.

6.5.4 Production of nested deletions using nucleases

Deletions can be generated fairly simply in double-stranded DNA by the action of several nucleases, such as *Bal* 31, exonuclease III and DNase I. Such deletions generated by the use of nucleases are often used to identify areas of DNA with regulatory functions. In such studies we often construct so-called nested deletions. During the production of such nested deletions progressively larger areas of DNA are removed from the target area, with a resultant loss of function. The method relies on being able to carefully control the activities of the nucleases being used so as to produce predictable deletions in the target DNA.

progressive
removal of
nucleotides

Several different nucleases can be used to delete sections of DNA and create nested sets of deletions (Figure 6.13).

Use Figure 6.13 to follow the description given below:

Bal 31 is an enzyme with both 5′ → 3′ and 3′ → 5′ exonuclease activity, and a weaker single-strand specific endonuclease activity. This enzyme produces blunt ended double-stranded DNA or short 5′ protrusions. To produce deletions, the target DNA is digested with a restriction enzyme that cuts at one end of the target sequence. The linear DNA molecule that is produced by this digestion is incubated with *Bal* 31 for different lengths of time. The termini are repaired with DNA polymerase, and the truncated regions of target DNA are isolated after digestion with a different restriction enzyme. This second restriction enzyme cuts the DNA at the end of the target DNA opposite to that cut by the first restriction enzyme. The truncated pieces of DNA can then be recloned.

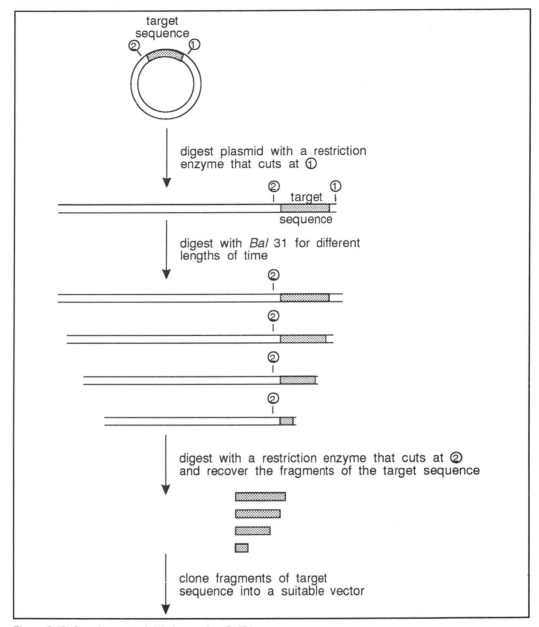

Figure 6.13 Creating nested deletions using *Bal* 31.

The enzyme exonuclease III (Figure 6.14) has $3' \rightarrow 5'$ exonuclease activity and $3'$ phosphatase activity. $5'$ mononucleotides are removed from the recessed or blunt $3'$ hydroxyl termini of double-stranded DNA. However, protruding $3'$ termini are completely resistant to the activity of this enzyme. Exonuclease III is often the enzyme of choice for the generation of nested deletions for several reasons. These are:

• it has a steady and predictable rate;

• the entire process can be carried out without intermediate purifications in one set of tubes (ease is an important factor for molecular biologists)!

- it does not exhibit any great sequence specificity;

- the activity of the enzyme is easily controlled by manipulating the concentration of sodium chloride;

- as protruding 3′ termini are completely resistant to the action of exonuclease III (as are 3′ termini with thionucleotide residues present), 'one-sided' or unidirectional deletions can be made.

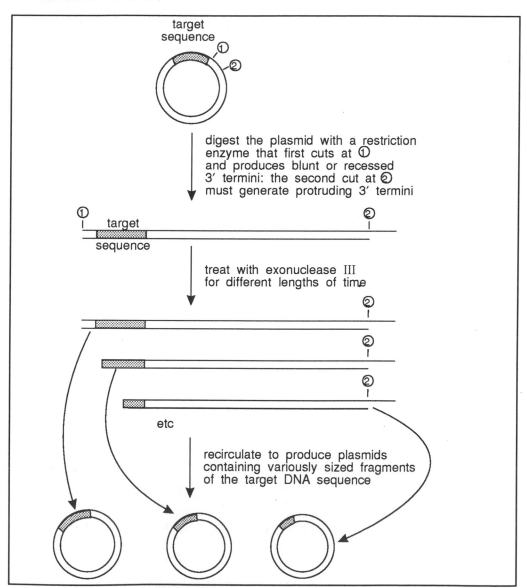

Figure 6.14 Creating nested deletions using exonuclease III.

To create deletions, double-stranded DNA is digested with two restriction enzymes, both with sites at one end of the target DNA sequence. The enzyme that cuts nearest the target DNA must produce either blunt ends or a recessed 3′ terminus and the other enzyme must produce a protruding 3′ terminus (or have a phosphorothiated nucleotide present) so as to be resistant to the action of exonuclease III. Digestion of the linearised DNA molecule then proceeds unidirectionally into the target DNA from the cleavage site of the first restriction enzyme. As with *Bal* 31, the termini can be repaired and the DNA recircularised.

The use of DNase I (see Figure 6.15) to cut double-stranded DNA has already been outlined earlier in this chapter. Also DNase I can be utilised to create nested deletions in DNA. First the double-stranded DNA is cut with DNase I at random sites. Subsequently the linear DNA is digested with a restriction enzyme which has a unique restriction site at one end of the target DNA sequence. Upon doing this the DNA between the site of DNase I cleavage and restriction enzyme cleavage is removed. As DNase I will cut the target DNA at random sites, the fragments of DNA that are removed can be of any size. After recircularisation the DNA can be used to transform host bacteria.

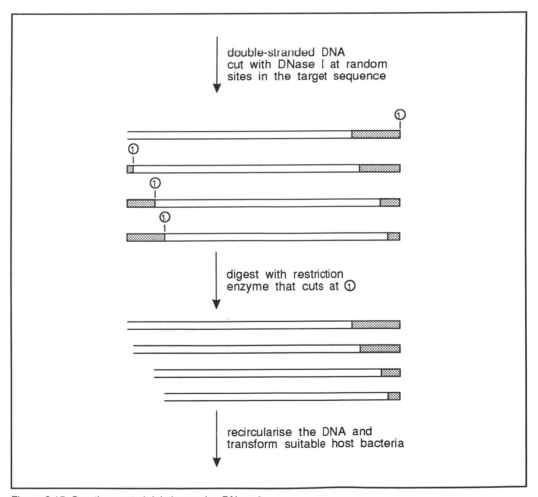

Figure 6.15 Creating nested deletions using DNase I.

SAQ 6.4

Select the most appropriate method from the list provided at the end of the SAQ to achieve the following objectives.

1) Randomly sited deletions need to be made in a section of cloned, double-stranded DNA. There are no convenient restriction endonuclease sites near or in the cloned target DNA.

2) Both randomly sited deletions, and randomly sited insertions also need to be made in the fragment of DNA described in 1).

3) A section of DNA known to possess a regulatory function has been cloned into a plasmid vector. The fragment contains no restriction sites. A method is needed that will locate, relatively precisely, any regulatory regions.

4) A similar piece of DNA as that described in 3) has been cloned into a plasmid vector. Restriction enzyme sites are available at each end of the cloned target DNA. A method is required to locate the approximate position of regulatory clones.

5) Deletions need to be produced in a cloned fragment of DNA. A restriction site at one end of the cloned fragment is available.

Methods available are:

Gap-sealing mutagenesis; linker-insertion mutagenesis; nested deletions with *Bal* 31; linker-scanning mutagenesis; nested deletions with DNase I, nested deletions with exonuclease III.

This was quite a difficult question so do not be disturbed if you did not get them all right. What this question will have shown you is that the battery of techniques that are now available for producing desirable mutations within a piece of DNA enables us to carry out desired mutational changes in a wide variety of genetic configurations. With experience you will become more proficient at identifying which particular technique is more likely to enable you to meet your objectives.

6.6 Random mutagenesis

The use of oligonucleotides or other methods to introduce one or a few point mutations is a well developed technique, but where many different point mutations are wanted in a target DNA sequence their use is too laborious.

Many random mutations can be produced by misincorporation of non-complementary nucleotides. For this purpose *Avian myeloblastosis* virus (AMV) reverse transcriptase is used. This enzyme is particularly error prone if forced to incorporate non-complementary nucleotides *in vitro* due to the omission of one dNTP from the reaction mixture. Moreover, AMV reverse transcriptase also lacks $3' \rightarrow 5'$ exonuclease activity so that errors that are introduced are not repaired *in vitro*.

AMV reverse
transcriptase

6.7 Use of *in vitro* mutagenesis

6.7.1 Studies of DNA functions

Mutagenesis has many academic and practical applications, for instance in the study of DNA regions controlling expression and in modifying proteins so as to introduce desirable characteristics.

identification of
regulatory
sequences

Many forms of mutagenesis *in vitro* have been applied to characterise areas of DNA that in some way regulate the expression of a gene. Mutagenesis can, for example, be used to isolate regions of DNA that control tissue specific expression.

In such investigations, large-scale deletions can be made in regions of DNA in the vicinity and upstream of the transcription initiation site. After mutagenesis, the mutated promoter can be used to drive the expression of a reporter gene, which enables a study of the pattern of expression. Mutation of the regulatory region should result in the loss of a particular pattern of expression.

intron/exon
borders

The location of such regulatory regions can be determined more precisely by the use of techniques such as linker-scanning mutagenesis. Similar studies can be used to study regions important in regulating RNA function such as intron/exon borders.

6.7.2 Protein engineering

Perhaps the major biotechnological application of mutagenesis *in vitro*, and particularly of oligonucleotide directed mutagenesis, is in protein engineering.

Proteins in which the sequence of amino acids have been altered in some way show great promise for a wide range of biotechnological applications. This is because by altering the amino acid sequence of a protein the characteristics of that protein can be altered. So, by engineering proteins it is thought that desirable characteristics can be introduced or enhanced. Such potentially useful modifications include increased thermal tolerance, different substrate specificity, increased reaction rates, resistance to inhibitors and activity over a wide range of pH values.

This is not to say though that protein engineering via oligonucleotide directed mutagenesis is easy. Before a protein can be engineered (at least with any hope of success) a great deal of information needs to be available about the structure and functioning of the protein. This is because it is easy to destroy the activity of a protein by changing the amino acid sequence, but very difficult to change the amino acid sequence and retain activity whilst introducing some desirable characteristic.

At present then, this need for information about the protein to be engineered is one of the main stumbling blocks for protein engineers as relatively little is known about many proteins.

SAQ 6.5

You are working for a large chemical company that produces biological washing powders. One problem with the current biological washing powder is inhibition of the protease, due to oxidation. This is thought to be specifically due to oxidation of a methionine residue of position 222. Using site directed mutagenesis this residue has been mutated to several other amino acids. You have been given some relevant data and have been asked to comment on which mutated protein would prove most useful.

Mutation	Relative reaction rate	K_M (x10^{-4} mol l^{-1})	Resistance to oxidation
wild-type	100	1.4	-
alanine	80	7.3	+++
cysteine	168	4.8	+

(reaction rates are relative to that of the wild-type, K_M is a measure of the affinity of the enzyme for its substrate).

Key: - = none; +++ = almost completely.

For use in a washing powder it is desirable that the protease will work at low substrate concentrations.

We will end this chapter by looking at some of the success stories of protein engineering.

6.7.3 Increasing thermal tolerance

Various approaches can be used to increase the thermal stability of proteins. One of the best studied examples is that of T4 lysozyme. Although T4 lysozyme is not a biotechnologically useful protein it is widely used as a model system for investigating the effects of changing the amino acid sequence of proteins.

Site directed mutagenesis has been used to convert an isoleucine (at position 3) to a cysteine. This change resulted in a disulphide bond forming with a previously unpaired cysteines. The mutated protein so obtained exhibited considerably increased thermal tolerance, including an increased half-life at increased temperatures and a resistance to both reversible and non-reversible thermal denaturation. It is not known precisely how this disulphide bond stabilises the protein at high temperatures.

One unfortunate lesson from this study, and the use of similar methods on other proteins, is that such methods may not be generally applicable. However, one amino acid substitution that does seem to be generally applicable to increase the thermal stability of proteins is the replacement of glycine residues with alanine.

6.7.4 Increasing specific activities

Interferon β is a protein which holds much pharmaceutical promise for the treatment of virus induced (and other) illnesses. Interferon β is produced by animal cells called fibroblasts and has a high antiviral activity. Recently, recombinant DNA technology interferon has also been produced in E. coli in order to increase production, but the activity of the purified protein was found to be approximately 100-fold less than that of interferon produced by animal cells. The cause was traced to incorrect formation of disulphide bridges between the three cysteine residues present. The cysteine at position

17 could be mutagenised to a serine, which produced a protein with full antiviral activity, which incidentally had increased stability during storage.

6.7.5 Increasing resistance to inhibition

α–1-antitrypsin is a member of a family of proteins known as serine protease inhibitors. It has an important role to play in preventing damage to lung tissue by neutrophil elastase (which causes emphysema). α-1-Anti-trypsin is susceptible to inhibition by oxidation of a methionine residue (residue 358) to methionine sulphoxide. This may be of particular importance to smokers, as smoking increases the amount of neutrophil elastase in lung tissue (due to increased numbers of leucocytes).

Leucocytes also release oxygen-free residues which can result in the oxidation of methionine residues to methionine sulphoxide. Thus, smokers who have to deal with increased amounts of neutrophil elastase, may have real problems due to decreased α-1-antitrypsin activity. However site directed mutagenesis can be used to replace the methionine residue at position 358 with valine. This creates an oxidation resistant α-1-antitrypsin that is still effective in inhibiting elastase. This modified protein may prove to be useful in treating emphysema by injecting patients with the modified protein.

Summary and objectives

In this chapter we have examined a wide variety of strategies for carrying out *in vitro* mutagenesis. We began by explaining some of the terminology associated with mutagenesis. We then described the overall strategy for chemical mutagenesis, oligonucleotide directed mutagenesis and techniques to introduce short deletions or insertions using nucleases. We explored the following techniques: single and double primer methods, gapped-duplex mutagenesis, coupled priming, the Kunkel method, gap-sealing mutagenesis, linker-insertion mutagenesis, linker-scanning mutagenesis and the production of nested deletions using nucleases. In the final part of the chapter, we briefly outlined the uses of *in vitro* mutagenesis.

Now that you have completed this chapter you should be able to:

- describe and identify from data, the various types of mutation that can be produced *in vitro*, such as insertions, deletions and point mutations;

- describe the use of chemicals to produce mutants *in vitro* and describe what type of mutation each chemical produces;

- describe and construct a flow diagram outlining the principles of oligonucleotide directed mutagenesis;

- describe several methods for oligonucleotide directed mutagenesis;

- describe *in vivo* and *in vitro* strand selection;

- describe several methods for producing large deletions or insertions in cloned fragments of DNA;

- describe the uses of mutagenesis *in vitro*.

Gene characterisation: mapping, sequencing and expression

Gene characterisation: mapping, sequencing and expression

7.1 Introduction

In this chapter we will begin to look at ways in which cloned DNA sequences can be characterised and identified.

We will start with a section on restriction mapping, a technique that can yield information on the site of the cloned sequence and the position of restriction sites for particular restriction enzymes. We then will focus on DNA sequencing, which can give us precise information on the base sequence in DNA. After having described these methods aimed at providing information on DNA structure we will continue with a description of the methods for characterising gene expression by assaying the mRNA or the protein produced upon expression.

7.2 Restriction mapping

physical
characterisation
of cloned DNA

Basic information about the 'physical characteristics' of a cloned DNA sequence is important for many manipulations carried out with cloned DNA sequences. Physical characteristics such as size of the cloned sequence, position of restriction sites and the sizes of fragments generated by digestion with particular restriction enzymes can be determined relatively simply by restriction mapping.

Several different methods can be used to determine the restriction sites of cloned DNA fragments. Here though we will only consider two of the more popular methods, double digestion and partial digests of end-labelled DNA.

7.2.1 Double digestion

This technique relies on the digestion of the DNA sequence of interest by two restriction enzymes. First the DNA is restricted with the individual enzymes and subsequently the fragments produced by enzyme 1 are treated with enzyme 2 and *vice versa*.

After digestion the DNA fragment with each individual enzyme, the fragments produced are analysed by agarose gel electrophoresis so that the size distribution of the fragments can be determined. Each of the fragments produced can then be eluted from the gel and then be digested by the other restriction enzyme. This produces a second set of fragments that again can be analysed by agarose gel electrophoresis. To complete the picture the DNA can be digested by both restriction enzymes simultaneously if conditions permit.

Figure 7.1 shows the results of agarose gel electrophoresis of the fragments produced by digesting a linear piece of DNA with either restriction enzyme 1 or restriction enzyme 2. As you know, each restriction enzyme cleaves the DNA at a particular sequence, called the recognition sequence. The fragments produced by the digestion of the DNA with a single restriction enzyme are separated on an agarose gel and stained so that they can be visualised.

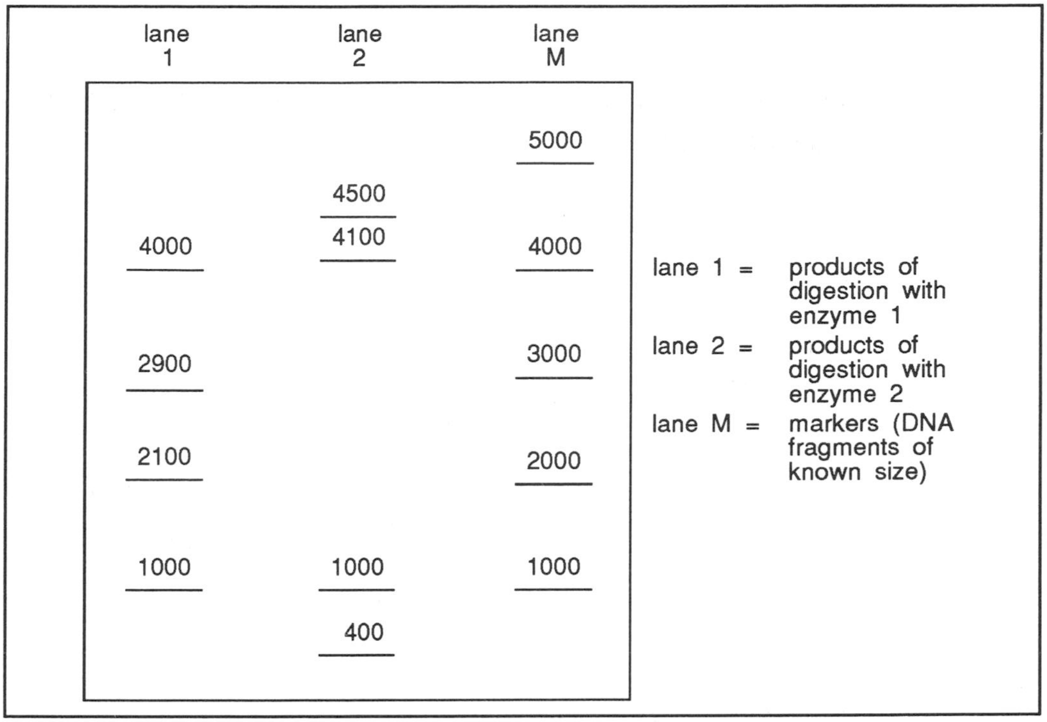

Figure 7.1 Agarose gel electrophoresis of the fragments produced by digestion with restriction enzyme 1 or 2. Numbers indicate the molecular weights of the fragments expressed in basepairs.

We can now use the information from the restriction enzyme digests to construct a restriction map of the cloned DNA sequence. We can follow it through in simple steps thus:

- Digestion with restriction enzyme 1 results in 4 fragments, the sizes of which are 1000 basepairs, 2100 basepairs, 2900 basepairs and 4000 basepairs. These we will call fragments A to D respectively.

- Digestion with restriction enzyme 2 results in 4 fragments, the sizes of which are 400 basepairs, 1000 basepairs, 4100 basepairs and 4500 basepairs. We will call these fragments E to H respectively.

After the data from the single digests have been assimilated we can proceed to the next step in the analysis.

- Each individual fragment is digested with the other restriction enzyme. That is fragments A, B, C, and D are digested with enzyme 2 and fragments E, F,G and H are digested with enzyme 1.

- The results of these second digestions are then subjected to agarose gel electrophoresis and give the following results.

Digestion of fragment A with enzyme 2 results in 2 fragments of 400 and 600 basepairs.

Digestion of fragment B with enzyme 2 results in 2 fragments of 1100 and 1000 basepairs.

Digestion of fragment C with enzyme 2 results in 1 fragment of 2900 basepairs.

Digestion of fragment D with enzyme 2 results in 2 fragments of 500 and 3500 basepairs.

Digestion of fragment E with enzyme 1 results in 1 fragment of 400 basepairs.

Digestion of fragment F with enzyme 1 results in 1 fragment of 1000 basepairs.

Digestion of fragment G with enzyme 1 results in 2 fragments of 600 and 3500 basepairs.

Digestion of fragment H with enzyme 1 results in 3 fragments of 500, 1100 and 2900 basepairs.

You will notice that when using this method we have several means at our disposal that can be used to check that the sizes of the fragments obtained in the second round of digests are correct. This may be important if, for instance, the second restriction digests are not efficient and result in partial digests. We have two simple methods to check for the correct fragment size distribution.

- The sum of the fragment sizes produced by digestion of each individual fragment with the second enzyme should equal the size of the individual fragment (remember that the sizes of the individual fragments A to H were determined by agarose gel electrophoresis). So we have for example:

  Fragment A (1000 basepairs) gives 2 fragments of 400 and 600 basepairs which is correct.

- The total of all the fragment lengths should add up to the total size of the cloned DNA sequence. This can be determined by adding together the fragment sizes obtained from the first digests and checking against the sizes determined by the second round of digests.

So let us use the results obtained with enzyme 1 which gives us fragments of 1000, 2100, 2900 and 4000 basepairs which gives a total size of 10 000 basepairs for the cloned sequence.

Checking this with the results of the second digests we have, using fragments A to D as examples, 400, 600, 1100, 1000, 2900, 500 and 3500 which gives a total of 10 000 basepairs, so all appears to be well.

We now have all of the information we can get from the results of the agarose gel electrophoresis experiments, so let us see what we can deduce from this information.

∏ First though can you list all the information?

1) The size of the cloned DNA sequence is 10 000 basepairs.

2) There are no sites for restriction enzyme 2 in fragment C. We know this because after digestion with restriction enzyme 2 the size of the fragment does not alter and only one fragment is present. Bear this information in mind for future use.

3)   Fragments E and F contain no site for restriction enzyme 1.

4)   Fragments A, B and D contain one site for restriction enzyme 2.

5)   Fragment G contains one site for restriction enzyme 1, fragment H contains 2 sites for restriction enzyme 1.

So let us produce a restriction map.

A good way to start is to look at fragment D, which we know contains one site for enzyme 2. This can be represented like:

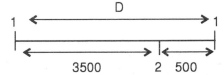

You can also see that a 3500 basepair fragment is produced by digestion of fragment G with enzyme 1. The two fragments thus overlap as shown below:

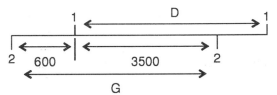

A 600 basepair fragment is also produced by digestion of fragment A with enzyme 2, again indicating the presence of an overlapping fragment. At the other end of the sequence where we have a 500 basepair fragment produced by digestion of fragment D we can see that digestion of fragment H also produces a 500 basepair fragment. The remainder of fragment H is occupied by a 2900 and a 1100 basepair fragment but at the moment we do not know which way round these are.

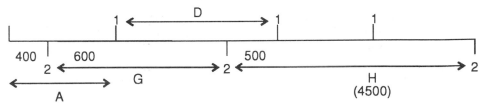

The order of these fragments and the complete map can be gauged from the results of digesting fragment B with enzyme 2. This produces a 1100 basepair fragment and a 1000 basepair fragment. There is then only one order in which these fragments can be arranged. This is illustrated below:

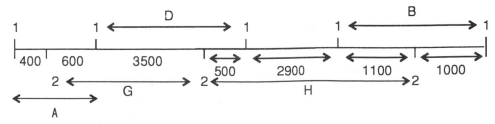

The 2900 fragment is fragment C.

The complete restriction map for this piece of DNA can be represented thus:

| 400 | 600 | 3500 | 500 | 2900 | 1100 | 1000 |
| 1 | 2 | 1 | 2 | 1 | 1 | 2 | 1 |

If you now look at SAQ 7.1 you will see how a complete map of a cloned DNA sequence can be built up. While you are working on this SAQ think of the factors that make the construction of a restriction map possible.

SAQ 7.1

Digestion of a 10 kb piece of linear DNA with the restriction enzyme *Eco* RI produces 4 products of 4200, 2800, 2000 and 1000 bp. Digestion with the restriction enzyme *Hin* dIII produces three fragments of 5000, 2600 and 2400 bp. Double digests were conducted. The results of which were:

digestion of the 4200 fragment produced two fragments of 3800 and 400 bp;

digestion of the 2800 fragment produced two fragments of 1600 and 1200 bp;

digestion of the 2000 fragment produced one fragment of 2000 bp;

digestion of the 1000 fragment produced one fragment of 1000 bp;

digestion of the 5000 fragment produced two fragments of 3800 and 1200 bp;

digestion of the 2600 fragment produced two fragments of 1600 and 1000 bp;

digestion of the 2400 fragment produced two fragments of 2000 and 400 bp.

Construct a map of this piece of DNA for *Eco* RI and *Hin* dIII.

7.2.2 Partial digests of end-labelled DNA

We can now go on to look at a more sophisticated way of producing restriction maps of cloned DNA. This approach combines partial digests of the cloned DNA with end-labelling.

polynucleotide kinase or Klenow enzyme

The cloned DNA fragment that is to be mapped is linearised and end-labelled with ^{32}P. This can be done in a variety of ways, for instance with polynucleotide kinase or by 'filling in' convenient restriction sites with the Klenow enzyme and labelled dNTPs.

The DNA is then partially digested with the chosen restriction enzyme. The conditions used are such that sometimes the chosen restriction enzyme fails to cut at a particular site, this generates a whole series of fragments that are the results of partially digesting the DNA. These fragments can then be separated by agarose gel electrophoresis and the fragments containing the label can be visualised by autoradiography of the dried gel and finally their sizes can be determined. A restriction map can then be determined by ordering the sizes of the fragments. The procedure is summarised in Figure 7.2.

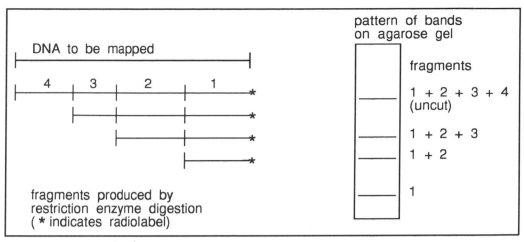

Figure 7.2 Restriction mapping using end-labelled DNA.

Obviously the examples you have seen are fairly simple. In particular the precise sizes of the DNA fragments produced by digestion with restriction enzymes are difficult to determine. In long pieces of DNA it is quite common for more than one restriction fragment to have (approximately) the same size. To some extent this problem can be overcome by quantifying the amount of DNA in the particular gel band. Small fragments of DNA can also be lost from the end of the gels during electrophoresis and although this fact would usually be detected it can still complicate analysis.

hybridisation of ovrlapping sequences

One further technique, nucleic acid hybridisation is sometimes used to confirm restriction maps. If restriction fragments are thought to overlap then they should also be capable of hybridising to each other as long as there is sufficient overlap for a stable hybrid to form.

The analysis of restriction enzyme maps can also be eased by the use of computer programmes.

Restriction mapping is a relatively quick and simple technique that can be used to determine approximately the sites for restriction enzymes in a piece of cloned DNA. It is often one of the first steps carried out in order to characterise a piece of cloned DNA. It does though produce only relatively limited data.

If more information about the cloned DNA is needed then the DNA may be sequenced.

7.3 DNA sequencing

DNA sequencing will give us precise information on the base sequence in DNA. We will look at three strategies for sequencing DNA:

- the chemical method (often called Maxam and Gilbert sequencing);

- the dideoxy method (often called Sanger's method);

- sequencing via RNA intermediates.

7.3.1 The chemical method

As its name suggests this method depends on the chemical cleavage of DNA to produce sequence data. This method utilises chemicals that break DNA after certain nucleotides. The DNA to be sequenced must be labelled so that after electrophoresis the fragments of DNA containing the label can be identified by autoradiography. Usually ^{32}P or ^{35}S is used to label the 5' end of the DNA. The DNA to be sequenced is made single-stranded and is divided into four samples. Each of the four samples is treated with a chemical that produces breaks in the DNA at specific bases. The conditions used are such that only limited cleavage at any one nucleotide occurs. This results in the production of a series of variously sized fragments, some of which contain the radiolabel.

end-labelled DNA

To achieve the desired cleavage specificity the bases must first be chemically modified in some way. These modification reactions are themselves quite selective, for example dimethylsulphate methylates guanine. The modified base is then removed from the sugar-phosphate backbone of DNA, usually by piperidine, and subsequently the strand is cleaved at the point lacking the base. These steps are outlined in Table 7.1.

Table 7.1 shows the chemical treatments that are used to induce breaks in the DNA at specific bases. Where A + G is listed as the base at which DNA is cleaved it indicates that the chemical treatment listed will cleave DNA at both A and G. A>>C indicates that the treatment is not absolutely specific, but DNA is cleaved at A residues much more efficiently than at C residues. A combination of treatments allows the complete DNA sequence to be determined.

Base at which DNA is cleaved	Base is modified with	Base removed with	DNA cleaved with
G	dimethylsulphate	piperidine	piperidine
A+G	piperidine formate	formic acid	piperidine
C+T	hydrazine	piperidine	piperidine
C	hydrazine + NaCl	piperidine	piperidine
A>>C	NaOH	piperidine	piperidine

Table 7.1 Chemicals used in DNA sequencing according to the chemical method.

∏   How do the chemicals mentioned in this table enable us to generate sequence data when some of the reactions are not specific for one base but cleave the DNA at two bases?

We can explain this in the following way.

Consider the sequence:

NNANGNNNNGNN   (N = nucleotide)

If we treat this with the G 'specific' reaction, we will obtain the following fragments:

GNNNNGNN   (8 nucleotides long)

    GNN   (3 nucleotides long)

If we treat with the A+G 'specific' reaction we will obtain the following fragments.

ANGNNNNGNN   (10 nucleotides long)

  GNNNNGNN   (8 nucleotides long)

      GNN   (3 nucleotides long)

If we compare these two sets of fragments, we can see that the 10 nucleotide fragment is unique to the A+G treatments (ie it does not appear in the G specific treatment. Thus this must be cleavaged at an A site.

prevention of DNA secondary structure formation in gels

The fragments generated in the four samples, upon treatment with the various chemicals, are resolved on a long polyacrylamide gel. Each of the four samples is loaded in a separate lane in the gel. The gel usually contains urea (at about $7 \text{ mol } l^{-1}$) to produce a gel with very high resolution, sufficient in fact to separate fragments that differ in length by only one nucleotide. The gel is usually run at a very high power setting so that a temperature of about 70°C is achieved in the gel to help ensure that the DNA does not exhibit any secondary structure. The urea present also helps to prevent secondary structures from forming. After electrophoresis is complete the fragments of DNA containing the radioactive label can be visualised by autoradiography.

A typical autoradiograph image (in diagrammatic form) of the results of sequencing by the chemical method is shown in Figure 7.3.

∏ Have you any idea what the sequence is and how it is read from the gel?

interpretation of gels

In order to read the sequence from the gel we must recall the explanation of the chemical cleavages given above. Furthermore we must realise that the template was labelled at the 5' end and that smaller fragments will have migrated further in the gel than larger fragments. We, therefore, can deduce the 5' → 3' sequence when we read the autoradiograph from the bottom of the gel upwards.

So we have, from the bottom of the gel shown in Figure 7.3, a band in both the C + T and C cleavage sample lanes. The base common to both is C, so the 5' base must be C. At the second position we have a band in the A + G lane, but no band in the G lane, therefore the base must have been an A. So the sequence so far is 5' CA. At the third position we also have a band in the A + G lane but also, one in the G only lane so the result is a G. The sequence now reads 5' CAG, and can be built up following the simple rules we have already met.

∏ Can you complete the sequence?

The complete sequence is 5' CAGCCGACCTTGAG 3'.

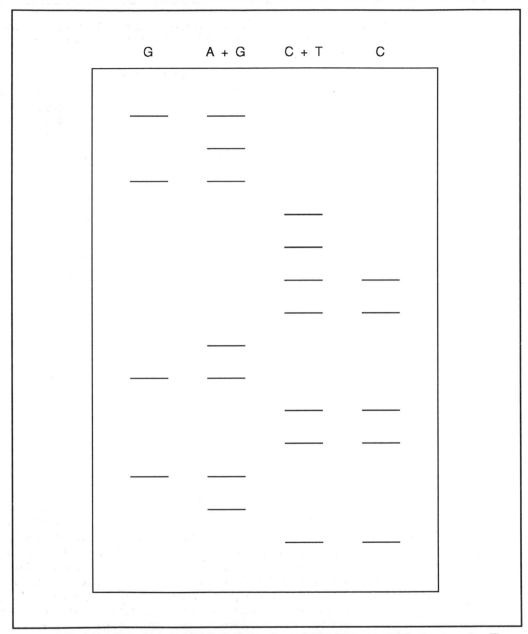

Figure 7.3 Typical pattern of bands seen on an autoradiograph after Maxam and Gilbert sequencing. The chemical cleavages are shown as lane headings.

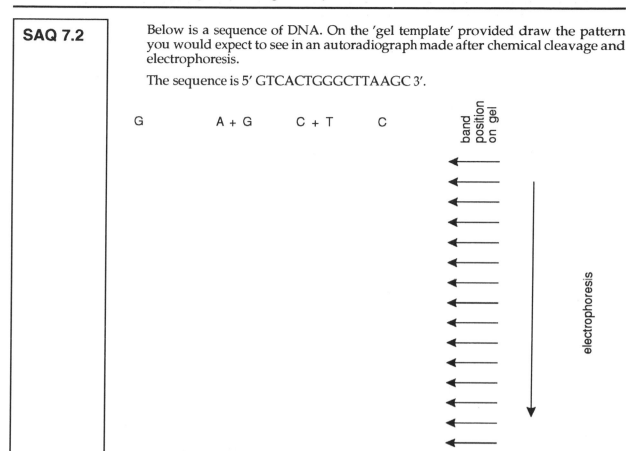

SAQ 7.2

Below is a sequence of DNA. On the 'gel template' provided draw the pattern you would expect to see in an autoradiograph made after chemical cleavage and electrophoresis.

The sequence is 5' GTCACTGGGCTTAAGC 3'.

7.3.2 The dideoxy method

inclusion of dideoxy-nucleosides primaturely terminates DNA polymerisation

In the dideoxy method we use a single strand of the DNA to be sequenced as a template for DNA synthesis catalysed by a DNA polymerase. In addition to the usual mix of four dNTPs, 2'3'-dideoxynucleoside triphosphates are added to the reaction mixture. When these base analogues are incorporated into a growing chain of DNA the polymerisation reaction will stop because the 3' ends of the analogues lack a hydroxyl group. These so called chain terminations are included at relatively low concentrations so that they will not be included at every possible position. In this way, a nested set of DNA fragments is produced that can again be resolved by gel electrophoresis. Also here the use of radioactive labels enables visualisation of the fragments on an autoradiograph of the gel.

In this enzymatic DNA sequencing the DNA to be sequenced is often, though not always, subcloned into a vector capable of producing single-stranded DNA. The vector also contains, next to the DNA which is to be sequenced, a sequence that can be used to initiate DNA synthesis by binding an oligonucleotide that will act as a primer. DNA synthesis, catalysed by a DNA polymerase, then occurs from this site and using the cloned DNA to be sequenced as a template.

enzymes used
for sequencing Several DNA polymerases can be used to carry out the sequencing reaction. Originally the Klenow enzyme was used but nowadays both natural and modified forms of T7 DNA polymerase and thermal tolerant DNA polymerases are used.

As in chemical sequencing we start with four samples. In the dideoxy method each reaction mixture contains one of the four possible dideoxy chain terminators in addition to the four usual dNTPs. At least one of the dNTPs is labelled with either ^{32}P or ^{35}S so that the fragments can be visualised by autoradiography. As an alternative to using labelled dNTPs, the primer oligonucleotide can be end-labelled and used to visualise the fragments (the primer oligonucleotide will of course be part of all the fragments produced by DNA polymerisation).

The DNA sequence can be read from the autoradiograph obtained after electrophoresis. Figure 7.4 shows a diagrammatic representation of such an autoradiograph. The pattern of bands can be read in the same way as that produced by the chemical sequencing method, although in the dideoxy method no two lanes will give bands at the same position. The DNA sequence in the example shown in Figure 7.4 reads as 5′ GGTCCAGACTTCGA 3′.

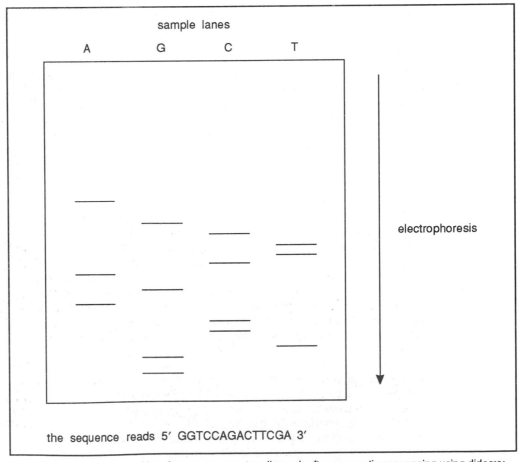

Figure 7.4 Typical pattern of bands seen as an autoradiograph after enzymatic sequencing using dideoxy chain terminators.

∏ Can you notice a difference between the sequence obtained by the chemical method and that obtained by the enzymatic method?

The sequence of fragments read from gels produced by enzymatic sequencing is actually the sequence of the complementary strand, whereas the sequence of the fragments produced by the chemical method is the actual sequence of the cloned DNA. This does not usually matter though as either strand of the original clone can be used as the template.

sources of
DNA used for
sequencing

Although sequencing using the enzymatic method is often carried out using single-stranded DNA produced from M13 type vectors, denatured plasmid DNA can also be used as a template for sequencing. Indeed, this method has many advantages, not the least of which is that there is no need to subclone the DNA to be sequenced into an M13 vector. The basic processes involved are the same as those applied to single-stranded template DNA. The double-stranded DNA can be denatured to provide a suitable template for sequencing by either linearising the plasmid followed by heat treatment or by treatment with alkali.

7.3.3 Sequencing via RNA intermediates

The enzymatic method of sequencing using dideoxy chain terminators can also be applied to sequencing DNA via RNA intermediates.

viral RNA
polymerase
promoters

In this method, the DNA to be sequenced is cloned into a vector that contains sequences or promoters for different viral RNA polymerases at each end of the cloned DNA fragment. Usually these promoters are for T3, T7 or SP6 RNA polymerases. Subsequently RNA copy is produced that extends through the cloned DNA and into the promoter at the opposite end. A DNA primer homologous to this opposite promoter sequence can then be used to prime enzymatic dideoxy sequencing reactions using reverse transcriptase as the polymerase. This method is illustrated below:

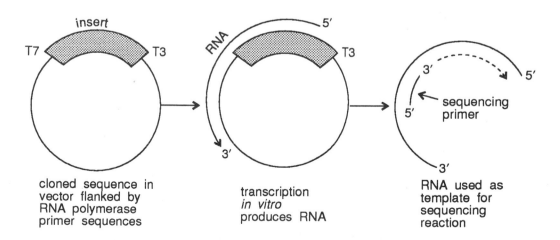

cloned sequence in vector flanked by RNA polymerase primer sequences

transcription *in vitro* produces RNA

RNA used as template for sequencing reaction

Nowadays the enzymatic approach to DNA sequencing is most commonly used due to its relative simplicity and speed compared with the chemical method. So, especially when large amounts of DNA are to be sequenced the enzymatic method is the one of choice.

The chemical method does though still find fairly widespread use today, and for some applications is the only method that can be used. One particular application is the localisation of methylcytosine and methyladenosine residues in mammalian and other DNA. Enzymatic sequencing cannot be used because data are obtained in terms of just the four bases used in the polymerisation reaction during formation of the complementary strand. By careful application of chemicals the position of naturally modified bases can be identified, for example methyladenosine reacts similarly to adenine sites in the A + G cleavage reaction but is much more resistant than adenine when the A>>C cleavage is used.

7.3.4 Automation of DNA sequencing

Manual sequencing using any of the currently available techniques is, in relation to the size of some projects, an expensive and time-consuming process. This is particularly so for the sequencing of the human genome, the total cost of the whole project using current technology would reach billions of dollars.

There is therefore considerable effort being put into developing sequencing protocols that are at least semi-automated. Two basic approaches to the problem of automation are being adopted; modification of individual steps within the usual framework of sequencing and the introduction of totally new techniques which eliminate some or all of the current steps in sequencing.

Some details of the emerging technologies are available, although many technologies are still at the experimental stage and it remains to be seen which will prove to be the most useful one.

use of
fluorescent
'tags'

One stage that is considered to be amenable to automation is electrophoresis of the fragments generated by the sequencing reaction followed by 'reading' of the fragments. Radiolabelling of the fragments can now be replaced by introducing fluorescent 'tags'. These tags can be detected using lasers to induce fluorescence. Each separate chain termination reaction can be labelled with a specific fluorescent tag. As the fragments are electrophoresed they pass by a laser and from the induced fluorescence it can be identified which base the tag corresponds to. This technique, and others like it, are hoped to produce a daily output of up to 15 000 bases.

Other more advanced techniques introduce more automation, for instance by introducing a 'robotic' system to carry out the actual sequencing reactions. A recent system that achieves an even higher degree of automation may result in a throughput of over 20 000 bases per day (and a cost reduction of at least 10 fold).

This is not, and is not intended to be, a complete examination of the automation of DNA sequencing. Currently, technology is moving very rapidly in this area and other more fundamental advances may become available shortly. However, the examples given do illustrate the directions that automation is taking.

7.4 Methods for characterising the expression of genes

In this section we will begin to look at the ways in which genes can be characterised in terms of expression. Such characterisation might include information on the size of the mRNA transcript produced, on the type of cells (or the area of the organism) the gene is expressed in and on the level at which the gene is expressed.

detection and
measurement
of specific
RNAs by
Northern
blotting

Basic data on all of these questions can be obtained by the use of RNA blotting and more specifically by using so called 'Northern Blots'. In addition to Northern blotting, which is routinely used in many investigations, other techniques are used in characterising gene expression. RNA titration analysis is used to measure absolute levels of mRNA, *in vitro* transcription is used to gather information on the rates of transcription and hybrid arrest and release of translation are used to correlate cloned DNA with proteins.

7.4.1 Northern blotting

Northern blotting basically consists of size fractionation of RNA followed by transferring the RNA to a solid support and the use of labelled probes to detect a particular sequence. By calibrating the fractionation step the size of the particular RNA species detected can be calculated. More target RNA present in any one sample will give a greater signal, and thus the method can be quantified to some degree. RNA can also be isolated from several different tissues so that expression patterns in whole organisms can be determined. Different developmental states can also be looked at in this way.

Northern blotting is therefore a simple and powerful technique regularly used to characterise gene transcripts. We can follow through the basic steps involved in Northern blotting by looking at Figure 7.5.

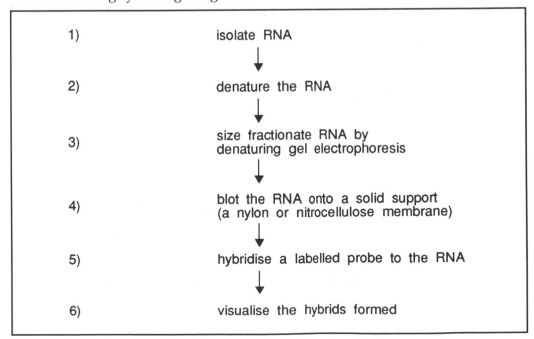

Figure 7.5 A flow chart illustrating the various steps involved in Northern blotting.

The RNA used for Northern analysis can be extracted by virtually any method. Either total or polyA enriched RNA can be used. Obviously more total RNA will be needed (as only about 1% of the total RNA will be messenger RNA) and 10 μg per sample lane is a common figure. If polyA RNA needs to be used, it can be isolated by chromatography through oligo dT cellulose or any similar material. The amounts of polyA RNA loaded on a gel can be much reduced compared with the amounts of RNA needed when using total RNA. The choice of whether to use RNA or polyA enriched RNA is dictated by a balance between the extra complexity involved in isolating polyA RNA and the greater sensitivity that this gives.

preparation of total and polyA enriched RNAs

The RNA to be size fractionated is then denatured. This step is used because, although RNA is considered to be single-stranded, it can have considerable secondary structure which can affect migration through the gel. Denaturing the RNA removes these secondary structures and results in the migration through a gel being largely a function of size.

RNA denaturation

RNA can be denatured via a variety of means, such as treatment with glyoxal, with a mix of formaldehyde and formamide or with methylmercuric hydroxide. Methylmercuric hydroxide is potentially lethal, and thus not attractive as a denaturing agent. Therefore only the use of glyoxal and formaldehyde/formamide will be discussed here.

Glyoxal is usually supplied as a 40% solution that must be deionised with an ion-exchange resin before use to remove impurities which form during storage. These impurities can destroy nucleic acids.

To denature the RNA, deionised glyoxal, sodium phosphate (at pH 7) and the RNA are mixed (sometimes dimethylsulphoxide is added as well), heated to 50°C for 1 hour and then left to cool. The denatured RNA can then be electrophoresed in agarose gels in sodium phosphate buffer which is recirculated to avoid the pH of the buffer rising above 8 as glyoxal readily dissociates from RNA above this point.

When using formaldehyde/formamide the agarose gel also contains a denaturing agent. The agarose gel in fact contains formaldehyde. The RNA sample is denatured by mixing deionised formamide, formaldehyde, buffer and RNA and heating to 55°C for 15 minutes.

The RNA is then electrophoresed and, if desired, can then be stained with ethidium bromide or more sensitively with acridine orange. However, staining the RNA reduces the efficiency of some subsequent steps. The gel can be calibrated by running denatured restriction fragments as size markers and staining only this particular lane.

The RNA can now be transferred (blotted) directly to the membrane, the apparatus used is illustrated in Figure 7.6.

nature of blotting material

Briefly the apparatus consists of a wick of about 3 sheets of a good quality filter paper (Whatman 3MM is commonly used) on top of which is placed the gel. All the exposed surfaces of the paper wick are covered with plastic film so as to stop the transfer buffer from passing past the gel rather than through it. The membrane (which is cut to size) can then be placed on top of the gel. For Northern blotting, nylon or chemically activated membranes are usually used. Any bubbles that have been trapped under the membrane have to be removed to ensure an even transfer. Further pieces of filter paper soaked in transfer buffer (cut to the same size as the gel) are then placed on top of the membrane and a pile (about 5cm high) of paper towels is placed on top of this. Transfer of the RNA to the membrane usually takes about 2 hours, but is often left overnight to ensure complete transfer.

The membrane is then removed, washed in transfer buffer to remove any traces of agarose that have adhered during transfer, blotted dry and then the RNA is fixed to the membrane by exposure to UV light or by baking in an oven.

The Northern blot is now ready for hybridising with the appropriate probes. The probes can be either DNA or RNA and can be radioactively labelled or be labelled so as to allow chromogenic or chemiluminescent detection.

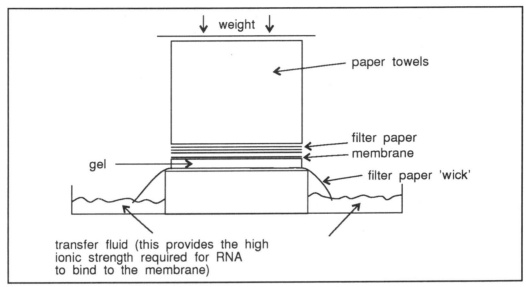

Figure 7.6 Blotting equipment used for Northern blots. The transfer fluid is usually 20 x SSC (SSC = standard saline citrate = 0.15 mol l^{-1} NaCl; 0.015 mol l^{-1} sodium citrate).

7.4.2 RNA titration analysis

Northern blotting is a powerful technique, its use for many investigations is routine. It does though have some limitations, especially when the amount of a particular RNA sequence has to be accurately determined. The drawback occurs because the investigator using a Northern blot is unable to ensure the total efficiency of transfer of nucleic acid from the gel to the membrane. The results from hybridisation experiments are also difficult to interpret due to the complicated kinetics associated with hybridisation to filter bound RNA.

RNA filtration is sensitive

RNA titration analysis does not require the use of any solid support such as a nylon membrane, and thus does not have many of the drawbacks associated with Northern blots. RNA titration analysis is the most accurate and sensitive method used to measure the absolute level of an mRNA species and can detect as little as 0.1 pg of a specific RNA.

The technique consists of adding excess amounts of the tracer, this is a probe that is homologous to the RNA of interest, to various amounts of cellular RNA. The hybridisation reaction is carried out in solution and is allowed to go to completion. Unhybridised tracer is removed by treating the sample with RNase and the number of hybrids formed between the tracer and cellular RNA is assayed. This can easily be measured by labelling the tracer to a known specific activity with a radiolabel. Under these conditions, if a graph of the amount of hybridised tracer is plotted against the amount of cellular RNA a straight line is produced. The slope of this straight line represents the percentage of the cellular RNA that is homologous to the tracer.

We can now go on to look at a more detailed analysis of the results. A graph of the type shown in Figure 7.7 should be obtained after an RNA titration analysis.

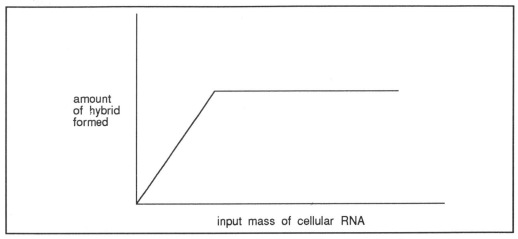

Figure 7.7 Graphical results of RNA titration analysis. The linear slope (positive gradient) indicates that the conditions used in the experiment were correct. The flat line represents conditions where the tracer is no longer in excess.

The fraction of cellular RNA homologous to the tracer can be calculated using the formula:

$$H = S \times L1/L2$$

where H is the fraction of RNA that is homologous to the tracer, S is the slope of the line obtained from a graph (as in Figure 7.7), L1 is the length of the mRNA being analysed and L2 is the length of the tracer.

The number of molecules of homologous RNA per cell can be calculated using:

$$N = (H \times R \times 6 \times 10^{23})/(F \times 339 \times L1)$$

where N = number of molecules of homologous RNA per cell, H is the fraction of RNA that is homologous to the tracer; R is the amount of RNA per cell, 6×10^{23} is the number of nucleotides per mole, F is the fraction of the total number of cells in the sample that express the gene, 339 is the average molecular weight of a ribonucleotide and L1 is the length of the RNA sequence being analysed.

RNA titration analysis gives us a very precise measure of the levels of a species of RNA in any particular sample, but in order for the result to be an accurate representation of the true cellular levels certain important criteria have to be met.

• The tracer must be in excess (5 fold). Initial estimates of the amount of RNA to be analysed are therefore necessary and this information can be gained from blotting experiments and comparison against known standards.

- The reaction must go to completion. In effect this usually means that reactions are left for at least 24 hours.

- The precise specific activity of the tracer must be known.

SAQ 7.3

You have been asked to analyse the results of an RNA titration analysis and give a value as to the number of molecules of a particular RNA that are present in one egg cell. A sample of several egg cells was used to provide the RNA. You have been given the following information:

the slope of the titration graph was 0.0001; the RNA of interest is 2000 bp long; the tracer RNA used was 200 bp long; the RNA content per egg cell is 2 ng.

7.4.3 Transcription *in vitro*

Although Northern blotting and RNA titration are powerful techniques they only give data as to the steady state levels of an mRNA species and are thus a measure of the final accumulation of the mRNA species.

The analysis of transcripts can be carried further by characterising the rate of transcription of certain target sequences. This technique allows the rates of transcription from several genes to be compared for instance, at a particular developmental stage. It of course also makes it possible to compare the transcription rates of one particular gene at different developmental stages.

application of transcription *in vitro*

Transcription *in vitro* is a technique that depends on the use of isolated nuclei from the tissue of interest. The isolated nuclei contain pre-initiated complexes (ie, transcription has begun) that are elongated *in vitro* in the presence of radio-labelled ribonucleotides. It appears that *in vitro* little new initiation occurs and thus the incorporation of radiolabel is thought to give a good indication of the number of active complexes that existed *in vivo* before the nuclei were extracted.

The labelled RNA can be added to cloned DNA sequences immobilised on membranes. Relative rates of transcription can then be gauged by quantifying the amount of RNA hybridised to the cloned DNA.

SAQ 7.4

Below are the basic stages involved in measuring transcription *in vitro*. Can you put them in the correct order?

1) hybridise to cloned DNA;

2) produce and label RNA *in vitro*;

3) isolate nuclei;

4) isolate RNA from contaminating DNA and proteins;

5) quantify the degree of hybridisation.

We can now look at each step in a little more detail.

Nuclei are usually isolated on what are called percoll gradients. Percoll is a suspension of specially coated silica. To do this cells are lysed in an iso-osmotic buffer that contains a non-ionic detergent. This type of buffer minimises damage to the nuclei during isolation, which is obviously an important factor if activity is required *in vitro*. The nuclei can then be partly purified by pelleting them during a slow speed centrifugation step. The nuclei from the pellet can then be further purified by centrifugation through a discontinuous percoll gradient. The gradient is often, though not always formed as shown in Figure 7.8. Nuclei are usually located after centrifugation at either the boundary between the 80% percoll and the 2 mol l^{-1} sucrose, or will pellet, according to their density (the density of nuclei can vary according to age and developmental condition).

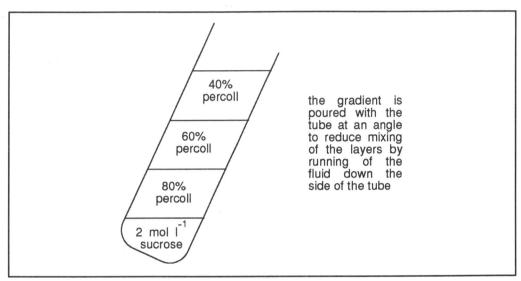

the gradient is poured with the tube at an angle to reduce mixing of the layers by running of the fluid down the side of the tube

Figure 7.8 Composition of a percoll gradient for isolation of nuclei. Note; different researchers use slightly different gradients.

contamination of nuclei

One potential problem with this type of gradient is that if the nuclei are found to pellet, they may be contaminated with starch granules which also pellet under this type of treatment. If this is the case the concentrations of the layers in the percoll gradient can be altered to try and separate the nuclei from the starch granules.

If all proceeds well, debris such as membrane fragments left over from the initial part of the purification, are usually found above the 40% percoll layer.

The area of the gradient containing the nuclei can be collected with a Pasteur pipette, the nuclei are washed several times to remove the percoll and then are resuspended in 50% glycerol. They can be stored at -80°C or below and are known to maintain activity (at least in some cases) for at least one year.

The isolated nuclei can then be used *in vitro* to produce labelled RNA. This is done by incubating the nuclei in a reaction mixture that has optimised conditions for RNA polymerase II activity. Also included in the reaction mixture are ATP, GTP, CTP and radiolabelled UTP. The extending RNA will incorporate the radiolabel. The radiolabelled RNA can be purified. This is usually done by centrifugation through a caesium chloride cushion. The caesium chloride and any unincorporated nucleotides can be removed by chromatography through sephadex and any contaminating polysaccharides removed by extraction with CTAB.

The RNA is then hybridised with either cloned DNA dotted onto membranes or DNA that has been Southern blotted. The results can be quantified by scintillation counting of the hybrids or by autoradiography and densitometric scanning.

7.4.4 Hybrid selection of mRNA and *in vitro* translation methods

Cloned DNA sequences can also be characterised by identifying the protein product of the gene. This is done by hybridising the cloned DNA to its mRNA and registrating the effect this has on translation. The procedure can proceed by either of two ways:

- the mRNA can be released from the hybrid after all the non-hybridising mRNAs have been washed away and then be translated *in vitro*;

- the formation of a hybrid can be used to inhibit translation. In this approach a protein that is found upon translation of total mRNA will be absent after hybrid formation.

Usually the cloned DNA sequence is immobilised on a membrane of nitrocellulose or chemically activated paper. Which type of membrane is used is a matter of choice as both have advantages and disadvantages. Nitrocellulose requires no special preparation and has a high binding capacity for DNA, however the DNA gradually washes off, so the same membrane can only be used a few times. Chemically activated paper, such as diazobenzyloxymethyl (DBM) paper needs special treatment just prior to use and has a lower binding capacity than nitrocellulose. However the DNA is covalently bound to the paper and so the membrane can be used repeatedly.

hybrid release translation
In hybrid release translation the membrane bound cloned DNA is hybridised with unfractionated mRNA or total RNA. The filter is then washed to remove unhybridised RNAs and the hybridised RNA is subsequently released from the hybrid by heating in a low salt concentration buffer. The RNA that is released can be used to drive translation *in vitro*. Translation *in vitro* uses a cell-free extract from either wheatgerm or rabbit reticulocytes and a labelled amino acid (often methionine with a radiolabelled sulphur). As translation proceeds the radiolabelled amino acid is incorporated in the growing protein. These labelled proteins can be analysed by polyacrylamide gel electrophoresis and autoradiography.

hybrid arrest of translation
Hybrid arrest of translation depends on the fact that an mRNA in a hybrid with DNA will not direct any synthesis of proteins in a cell free system. Thus, when the RNA is used in a cell-free translation system, hybridisation of a specific mRNA to the cloned DNA will result in a pattern of protein bands in which one protein is absent that was present in the pattern observed when total mRNA is translated.

Both hybrid release and hybrid arrest of translation allow cloned DNA to be correlated with certain proteins produced by the cell, and provide invaluable evidence concerning the nature of the cloned DNA.

Further proof for the identity of the protein product of a particular cloned DNA sequence may be obtained from heterologous gene expression experiments.

Basically, the cloned DNA sequence is used to transform either bacteria, lower eukaryotes such as yeast, or mammalian cells. The protein so produced can be purified if required and biological activity can be measured.

It may also be possible to assay the biological activity *in vivo* if that activity was not present in the untransformed cells. Alternatively immunological techniques can be used to confirm the identity of the protein.

Whether bacteria or eukaryotic cells are used depends on many questions. Use of bacteria for instance often allows a relatively simple purification of the protein, although it has the disadvantage that the proteins may not be correctly modified (phosphorylation and glycosylation may not be correct) and may thus lack biological activity. Transformation of eukaryotic cells is technically more demanding but the proteins produced are likely to be biologically active, and also some idea of sub-cellular localisation may be gained.

7.4.5 Hybridisation *in situ*

In this section we will look very briefly at techniques for localising gene expression by hybridisation *in situ*. Hybridisation *in situ* is a very complicated technique, and several alternatives to many, if not all, of the individual steps are available. Thus here we will look at an overview of the technique and indicate why and how it is used.

A generalised flow chart for hybridisation *in situ* is shown below:

the tissue to be used for hybridisation
is fixed, dehydrated and embedded

↓

a section of the tissue is mounted
on a prepared microscope slide

↓

the tissue section is prepared for
hybridisation

↓

hybridisation is carried out and the
results are visualised

We can now look at each step in a little more detail.

The aims of the fixation, dehydration and embedding are basically two fold. Firstly the morphology of the tissue has to be preserved if a meaningful insight about the type of cell where expression occurs is to be gained and secondly we have to ensure that the RNA is retained within the tissue during subsequent steps.

.agents use as fixatives

A variety of agents such as glutaraldehyde or a mixture of formalin, alcohol and acetic acid (FAA) are often used as fixatives. The tissue can be dehydrated by immersion in increasing concentrations of alcohol (such as ethanol, although other alcohols can be used) and embedded by melting paraffin wax around the tissue, allowing it to infuse into the tissue and then to set. This allows cutting of very thin sections of the tissue that can be examined under a microscope.

removal of proteins

When the tissue has been mounted, the pre-hybridisation treatments are begun, the aim of which is to allow efficient access of the probe to the target RNA and to reduce the amount of non-specific binding of the probe to the tissue. Treatments usually include limited removal of proteins with either pronase or proteinase K to increase accessibility and washes with BSA (bovine serum albumin) to reduce non-specific probe binding.

use of
ani-sense RNA
to allow
removal of
non-hybrids

The tissue can then be used for hybridisation. Single-stranded (anti-sense) RNA probes produced *in vitro* are often used because they allow removal of nonhybrids by treatment with RNase. Moreover RNA-RNA hybrids have a high thermal stability which allows the use of high temperatures in subsequent wash steps, which increases stringency and reduces non-specific binding. The actual hybridisation occurs on the tissue sample, usually under a cover slip, with the whole assembly immersed under mineral oil.

After hybridisation the tissue sections are washed with RNase and low-salt concentration buffers at high temperature to reduce the amount of non-specific probe binding. The slides are then dipped in liquid photographic emulsion and left. The emulsion can then be developed and the tissue examined by microscopy to localise the hybridisation signal.

Hybridisation *in situ* provides a way of assaying gene expression in a particular organ or cell type when it is impossible to separate it from other structures or cell types.

SAQ 7.5

You have been given several tasks, which are listed below. From what you have learned, choose from the methods listed, the method that you think is most likely to give results.

Tasks:

1) you need to identify the protein product of a cloned cDNA;

2) you need to know where in a plant a particular RNA sequence is expressed;

3) you need to know if the different levels of an mRNA found by Northern blotting are due to different transcription rates in different parts of a plant;

4) you need to sub-clone small, easily sequenced pieces of a large cDNA into a vector, and need to know the location of restriction sites;

5) you need to know the precise number of RNA molecules in a particular cell type;

6) as a precursor to 5) you need to know what percentage of cells in a mixed sample is expressing the particular sequence;

7) you need to pinpoint the location of several restriction sites;

8) you need to locate the position of methylated bases in a piece of DNA.

Methods:

a) transcription *in vivo*;

b) RNA titration analysis;

c) hybrid arrest of translation;

d) Northern blotting;

e) hybridisation *in situ*;

f) sequencing using dideoxy chain terminators;

g) restriction mapping the DNA;

h) chemical sequencing (Maxam and Gilbert).

Summary and objectives

In this chapter we have dealt with methods to characterise cloned DNA sequences. We started with a section on restriction mapping followed by a section on DNA sequencing. In the final section methods for characterising gene expression were described.

After reading this chapter you should be able to explain ways in which cloned DNA sequences can be characterised, both physically and in terms of expression.

You should be able to:

- explain how data about the restriction sites present in a piece of DNA can be obtained;

- explain how more detailed information about the physical characteristics of a piece of DNA can be obtained by sequencing;

- describe several different methods used for sequencing;

- explain how the expression of a gene can be characterised;

- describe the use of Northern blots;

- describe how RNA titration analysis works;

- describe transcription *in vitro*;

- describe how a cloned sequence can be characterised in terms of its protein product;

- describe briefly how hybridisation *in situ* works and what its uses are.

Screening libraries: identifying genes and regulatory sequences

Screening libraries: identifying genes and regulatory sequences

8.1 Introduction

In this chapter we will continue to look at ways in which cloned DNA sequences can be studied. In the previous chapter we have seen how cloned DNA sequences can be characterised physically by restriction mapping and sequencing. We have also seen that the cloned DNA can be characterised in terms of gene expression by assaying for mRNA and /or protein products. In the present chapter, we will look from a slightly different angle at the characterisation of the 'structure' of cloned DNA sequences. Most techniques outlined here are aimed at providing information about structural features of cloned genomic DNA in order to identify important regions such as 'promoters'. Such sequences can of course be involved in gene expression, but their expressions cannot be directly assayed by measuring the mRNA or protein product.

We will first remind you, briefly, of how clones can be isolated and how a gene can be localised within the cloned DNA. We will deal with the methods that can be used to locate the ends of genomic sequences transcribed into RNA and subsequently we will describe the chromosome walking technique. This technique, that is also called overlap hybridisation, is a technique for isolating overlapping recombinants in order to 'walk' from one position on the chromosome to another nearby position.

We will then focus on the techniques used in analysing regulatory regions of the DNA and finally will describe methods for studying protein/DNA interactions.

8.2 The isolation and initial characterisation of a clone

For the purpose of this chapter we will assume that a genomic library containing the relevant clones is available (the techniques used in the construction of genomic libraries have been outlined in Chapter 5). Obviously, in order to characterise a clone it is necessary to isolate it from the library and furthermore it may be necessary to localise the sequence of interest within the cloned DNA sequence.

Methods for isolating recombinant phage λ from genomic libraries have been dealt with already and therefore we will only briefly summarise the processes included here.

The method most widely used to identify recombinant phage λ is plaque hybridisation. The plaques produced can be screened with a suitable probe (often a DNA sequence) and a plaque which gives a positive result can be isolated. The recombinant phage can then be grown again and rescreened to check that no 'contaminating' phage are present. If the re-screen is successful and a pure phage clone can be isolated then the DNA from this phage can be isolated and characterisation can be started.

We have already seen that initial characterisation by restriction mapping is very useful, and this can be even more so when we are dealing with genomic DNA clones, which can be of a very large size. A restriction map of a cloned piece of genomic DNA will prove to be a very useful starting point for many subsequent manipulations.

8.2.1 Localisation of the 'gene' within the cloned DNA

Several methods can be used to determine the position of a gene within cloned DNA. The general location of a gene within a cloned DNA can be determined by either blotting RNA and probing this blot with labelled fragments of the cloned DNA, or alternatively fragments of the cloned DNA can be separated by agarose gel electrophoresis, blotted and probed with labelled RNA or cDNA.

localisation by blotting techniques

Both of these methods are fairly simple and yet fairly sensitive, and can yield other information. For instance, blotting of RNA will give information about the size of the transcripts. Furthermore, we can also deduce from the data obtained whether any other 'genes' are present in the cloned DNA. Obviously the more precise the available restriction map of the cloned DNA is, the more accurately the position of the gene within the cloned DNA can be determined.

Other, more elaborate methods are also available for determining the position of a gene in a cloned DNA sequence.

R-loop analysis

One method is the so called 'R-loop' analysis, which depends on the relative strength of RNA-DNA hybrids compared with DNA-DNA hybrids under certain conditions. If double-stranded DNA, such as phage DNA, that contains a gene of interest is incubated with mRNA at high temperatures in the presence of formamide, the mRNA can displace one strand of the DNA as it hybridises to the complementary strand. The displaced, non-complementary, strand of DNA forms an 'R-loop' which can be visualised by the use of an electron microscope. This R-loop structure can thus be used to mark the position of a gene in the DNA.

Using these techniques, we can locate, quite accurately if a comprehensive restriction map is available, the position of a gene within a piece of cloned DNA. We may though still need more detailed information, such as where the 3' and 5' ends of the gene are. We will now look at two methods for locating the ends of genomic sequences transcribed into RNA, namely nuclease protection assays and primer extension.

8.2.2 Nuclease protection

In nuclease protection assays, single-stranded end-labelled DNA is hybridised to complementary RNA and then the hybrid is incubated with specific nucleases which hydrolyses single-stranded nucleic acids. Finally, the resulting hybrid is analysed using gel electrophoresis and autoradiography. The technique enables us to locate the ends of a gene. Figure 8.1 summarises the different stages of the process.

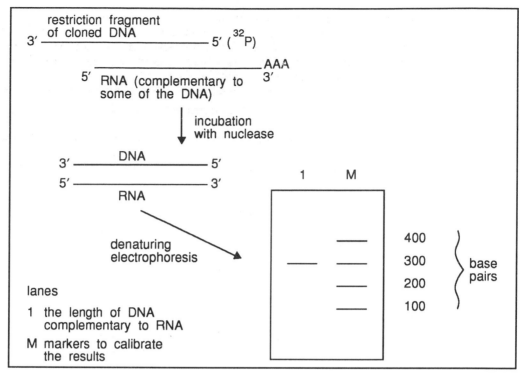

Figure 8.1 Nuclease protection assays. The approximate distance between the termini of a gene can be estimated following autoradiography of hybrids separated by denaturing gel electrophoresis.

Let us now consider nuclease protection in some more detail. The stages are:

- a restriction fragment of the cloned DNA is made single-stranded and is end-labelled, usually with ^{32}P;

- the labelled DNA is hybridised to RNA that is complementary to part of the DNA;

- the mix produced after hybridisation is incubated with a single-strand specific nuclease;

- the hybrid obtained after nuclease treatment is analysed by denaturing gel electrophoresis followed by autoradiography.

determination of the gene terminus on a DNA fragment

Ideally we would expect to see a single band on the autoradiograph made after electrophoresis. This band can be 'sized' by comparison with known standards. We then know what length of DNA was protected from digestion by the single-strand specific nuclease because the complementary RNA was hybridised to it. By comparing this size to the size of the restriction fragment we started with, we can deduce how far the terminus of the gene was removed from the end of the restriction fragment. Obviously, by using smaller pieces of DNA, we can obtain more precise information. A detailed restriction map is, therefore, a good starting point because it allows the isolation of fragments of the cloned DNA.

Let us now give some additional comments on aspects of the nuclease protection assay.

Some initial characterisation of the cloned DNA is necessary. It must have been restriction-mapped and the orientation and approximate location of transcribed regions should be known (see above). The DNA must also be made single-stranded and has to be labelled, usually with ^{32}P, to allow visualisation of the products by autoradiography.

The radiolabel must of course always form part of the RNA-DNA hybrid and must not be removed by the nuclease treatment. We, therefore, label the DNA on the 5' end if the 3' end of the gene (that will of course hybridise to the 5' terminus of the RNA) is to be mapped. If the 5' end of the DNA is to be mapped, the 3' end of the DNA must be labelled. This principle is illustrated in Figure 8.2

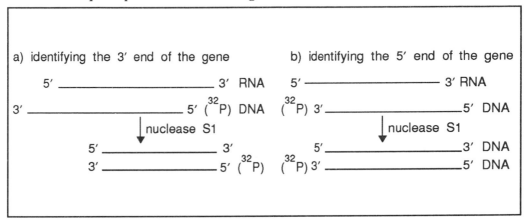

Figure 8.2 Mapping strategies using nuclease protection assays.

nuclease S1 digestion of SS nucleic acids

The most frequently used (and probably the least expensive) nuclease is S1 nuclease. S1 nuclease digests single-stranded nucleic acids much more rapidly than double-stranded nucleic acids when high salt concentrations are used and the experiment is carried out at low temperature. Another nuclease, called mung bean nuclease, has similar properties and can be used instead of S1 nuclease.

exonuclease VII

A special nuclease from *E. coli*, called exonuclease VII also finds application in nuclease protection assays. Exonuclease VII requires free 5' and/or 3' ends to start digestion; nuclease S1 does not need free ends to start digestion. The differences between these two types of nuclease are illustrated in Figure 8.3

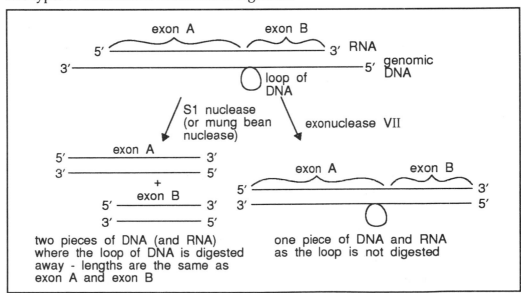

Figure 8.3 A comparison of the activities of S1 nuclease (or mung bean nuclease) and exonuclease VII in nuclease protection assays.

It will be evident from Figure 8.3 that upon comparison of the results obtained by digestion with nuclease S1 (or mung bean nuclease) with those obtained by digestion with exonuclease VII, information about the lengths of both introns and exons in cloned DNA can be obtained.

The labelled DNA should be present in a molar excess over the complementary RNA. This can be checked by carrying out control hybridisations with different amounts of RNA; the intensity of the signal generated should be proportional to the amount of RNA present.

Now we have seen how nuclease protection assays work let us look at a question.

SAQ 8.1

1) You want to map the 5′ terminus of an RNA. Which end of the DNA do you label? The experiment is illustrated below.

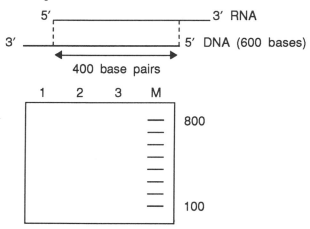

2) Draw on the gel:

   a) in lane 1 the band size you would expect if the probe was incubated without nuclease S1 digestion and then electrophoresed;

   b) in lane 2 the band size you would expect if the probe was digested with nuclease S1 in the absence of RNA;

   c) in lane 3 the results of a full nuclease S1 protection assay.

We can now go on to look at an alternative method used for determining gene termini.

8.2.3 Primer extension

In principle, primer extension is quite the opposite of nuclease protection as it relies on polymerisation rather than on digestion. In essence, a radioactively labelled sequence of DNA entirely from within the gene is hybridised to complementary RNA and extended using reverse transcriptase. The primer extension product is thus longer than the DNA sequence used to start the process. The extension reaction halts when the end of the RNA is reached. The primer extension procedure allows the terminus of the RNA to be mapped.

The technique is summarised in Figure 8.4.

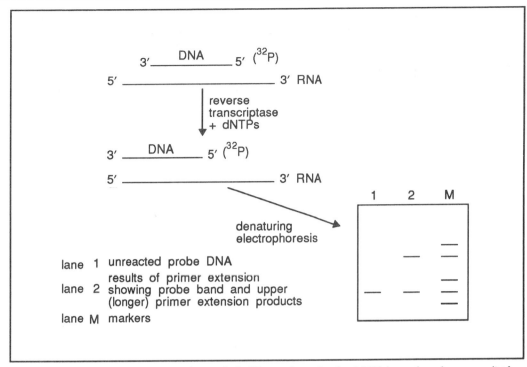

Figure 8.4 An outline of primer extension analysis. The newly synthesised DNA is produced as a result of extension from the primer.

The DNA used in primer extension is usually a restriction fragment of cloned DNA, and usually comes from near the 5′ terminus. The fragment is often generated by digestion with two separate restriction enzymes so that both ends of the DNA molecule will have different base sequences and thus circularisation will not occur. This allows single-stranded DNA to be isolated by denaturing gel electrophoresis.

The RNA used can be either polyA RNA or total cellular RNA. Total cellular RNA is obviously easier to isolate but can lead to the production of many prematurely terminated fragments if the distance from the DNA primer to the 5′ terminus is greater than about 200 nucleotides.

The hybridisation conditions used need to be carefully controlled. Optimum hybridisation temperatures for a given set of conditions depend on the length of the DNA and its composition. The DNA should be present in excess, although not drastically so, otherwise non-specific priming can occur.

primer
extension
sequenced by
Maxam and
Gilbert method

A great advantage of primer extension is that the primer extension product can easily be sequenced after being recovered from the gel, for instance by using Maxam and Gilbert chemical sequencing. This allows precise localisation of the 5′ terminus. However, for sequencing to be carried out, the process has to be scaled-up so as to provide sufficient radioactively labelled material for sequencing.

<table>
<tr><td>

SAQ 8.2

</td><td>

You have been asked to map the 5′ terminus of an RNA by primer extension. The experiment is illustrated below.

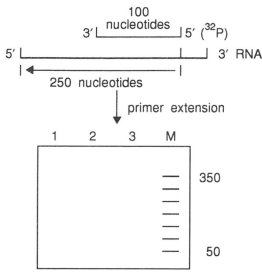

Draw on the gel the bands you would expect:

1) in lane 1 if the unreacted probe DNA were electrophoresed;

2) in lane 2 if the probe DNA was added to a reaction mixture in which no complementary RNA was present;

3) in lane 3 if the primer extension reaction was successful.

</td></tr>
</table>

You may of course find after some of these techniques that the complete coding sequence of the gene, or some other area of interest such as a promoter is not present in the piece of cloned DNA. If this is the case it may be possible to isolate full length genes by a technique known as chromosome walking.

8.3 Chromosome walking

Chromosome walking is a technique for isolating clones containing overlapping DNA sequences from a genomic library in order to 'walk' from a position on a cloned and analysed gene to another position nearby on the chromosome. The technique is outlined in Figure 8.5.

Chromosome walking starts with the isolation of a terminal fragment from the starting clone. This terminal fragment is labelled and used as a hybridisation probe (probe 1 in Figure 8.5) for scanning a genomic library. It will identify all clones in the library that contain the fragment. These DNA clones are then restriction mapped to identify the clones that overlap with the probe and extend into the region adjacent to the DNA fragment in the starting clone. Then the terminal fragment of this newly isolated clone is isolated, labelled (probe 2 in Figure 8.5) and used to start the second 'step' in the walk. The process can be continued until the full length of a gene, or even sets of genes, have been isolated. In Figure 8.5 all 'steps' go to the right, but of course the walk can be extended in both directions.

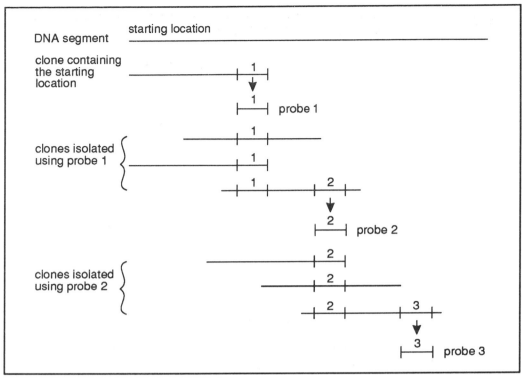

Figure 8.5 An overview of the chromosome walking technique (see text for details).

The steps involved in chromosome walking are thus as follows:

- screen genomic library with a labelled piece of DNA prepared from the initial clone;

- isolate clones that give positive hybridisation signals, analyse these by restriction digestion and identify homologous fragments by Southern blotting;

- identify overlapping fragments from the DNA inserts in the isolated clones;

- prepare a labelled probe from some of the overlapping fragments and repeat the procedure.

size of steps depend on the vectors used

The libraries used for chromosome walking are usually constructed in phage vectors or in cosmids. Both have advantages and disadvantages, and therefore, the choice will depend on the particular situation. Phage λ vectors generally contain smaller DNA inserts (up to about 20 000 basepairs) than do cosmids (these contain up to about 40 000 basepairs). Thus we have to take smaller 'steps' when using phage libraries. However, phage libraries are generally easier to construct, store and work with. Whichever vector is chosen, it is vitally important that the genomic library used is 'complete' so as to avoid situations where the walk has to be abandoned because a clone is missing. It is also an advantage if the inserted DNA can easily be removed from the rest of the vector DNA, so that screening of the library is not complicated by labelled DNA hybridising to vector DNA during plaque hybridisations.

One major problem that may be difficult to overcome is the presence of repetitious DNA sequences. If probes containing repeated DNA sequences are used, the walk will 'split' into several different directions due to hybridisation to other repetitive DNA sequences in otherwise unrelated DNA sequences.

Now that we can isolate and characterise full-length genes, and go on to isolate sequences flanking the coding region of genes, let us look at how potentially very interesting and useful regulatory sequences can be investigated. We shall consider several methods that can be used to investigate transcriptional regulatory elements such as promoters and enhancers. Many of these techniques are used in combination with *in vitro* mutagenesis as was outlined in Chapter 6.

8.4 Analysis of regulatory regions

Regulatory sequences are usually considered to be situated in the flanking regions of genes, although this may not always be the case as some regulatory sequences have been found in intragenic regions.

One approach that can be applied to studying regulatory regions situated in both flanking regions and in intragenic sequences is to transform cultured cells with a genomic DNA fragment that contains both coding and flanking regions. If the mRNA produced from the introduced gene is of a sufficiently different sequence or size to any related sequences produced by the cells, then mRNA levels can be studied using Northern blotting. By using deletion and site-directed point mutagenesis, regulatory regions can be identified by assaying the effect of mutagenesis on the amounts of mRNA present (see also Chapter 6). This technique has the advantage that regulatory regions in both the flanking and intragenic sequences can be identified. The technique does though have several limitations:

* the gene must be relatively small (say less than 10 000 basepairs) for all the manipulations to succeed;

* we must be able to distinguish the mRNA produced from the introduced gene from the endogenous RNA produced by the cultured cells that are used for transformation.

If the second requirement is not fulfilled, the RNA produced by the gene introduced by transformation can be marked in another way. This can be done by the introduction into the gene of a synthetic nucleotide sequence that will be transcribed or by the introduction of a plasmid sequence that can be detected by Northern blotting.

minigenes

The problem of gene size is of course not solved by this introduction of new DNA sequences, and if the gene is too large to be manipulated a different technique will have to be used. One approach used is to delete some internal exons from the gene to reduce its size to manageable proportions. These constructs are often referred to as 'minigenes' and have the added advantage that RNA produced from the minigene can be distinguished from any endogenous RNA of a similar nature by virtue of size. Minigenes do though have several disadvantages. Deletions from the gene cause alterations in the transcript that is produced which may lead to the transcript becoming unstable (perhaps due to alterations in secondary structure) and may thus give a false negative result when assayed. The deletions may also cause the loss of some regulatory regions, which as mentioned before, can sometimes be located in intragenic sequences.

8.4.1 Reporter genes

These methods described in the previous section depend on measuring the transcripts from complete or modified genes. There are methods, however, that are becoming increasingly popular in both animal and plant biology, which depend on measuring transcription or the products of transcription of a quite different gene which is linked to the putative regulatory regions of the gene. This is called the use of reporter genes.

globin gene used as a reporter gene

Reporter genes have a coding region which can easily be joined to the regulatory sequences of other genes. Reporter genes can be of several types. Eukaryotic genes such as globin genes have been used as marker genes because transcription is usually restricted to one tissue. Therefore other cell types can be transformed with a globin coding sequence linked to a putative promoter sequence without endogenous globin transcripts interfering with the assay.

reporter genes coding for easily assayed enzymes

Other commonly used reporter genes are bacterial genes that code for enzymes that are easily assayed. Such genes are the chloramphenicol acetyl transferase gene, the neomycin phosphotransferase (II) gene, the β-galactosidase gene, the β-glucuronidase gene and the luciferase gene (this last mentioned gene is not a bacterial gene but is included here because it can be used in the same way as the other listed genes).

chloramphenicol acetyl transferase

The chloramphenicol acetyl transferase gene is probably the most widely used reporter gene in mammalian, prokaryotic and plant cells. The assay for chloramphenicol acetyl transferase is based on the acetylation of ^{14}C-chloramphenicol and the subsequent analysis of the products by thin-layer chromatography on silica TLC plates. The acetylated ^{14}C-chloramphenicol can be visualised by autoradiography. Chloramphenicol can be acetylated in more than one position so multiple autoradiography 'spots' may be seen. To get exact information about the amount of chloramphenicol acetylated, the 'spots' on the TLC plate can be extracted with solvents and counted in a liquid scintillation counter. The activity of chloramphenicol acetyl transferase can be inhibited by some oxidases which can be a particular problem in some plant species.

neomycin phospho- transferase

Neomycin phosphotransferase is usually assayed by incubating ^{32}P-ATP with kanamycin (a substrate) and measuring the radioactivity of the products formed in a scintillation counter. Assaying neomycin phosphotransferase in plants may again present problems as non-specific phosphorylation can occur. A different, more complex assay then needs to be used. In this assay system the constituents are separated by the use of non-denaturing polyacrylamide gels, which are then exposed to kanamycin and ^{32}P-ATP, after which the products are blotted onto special paper and counted or assessed by size in comparison with known phosphorylated products.

β-galactosidase

β-galactosidase can be assayed in a gel system like the one used for neomycin phosophotransferase assays. It is often necessary to use this system in plants because they contain endogenous β-galactosidases that need to be separated out from the reporter gene product. Here though the gel can be stained with 4-methylumbelliferol - released from 4 methylumbelliferyl-β-D-galactoside by β-galactosidase. The product released by β-galactosidase activity fluoresces under UV light.

Alternatively, we can assay β-galactosides in crude cell extracts spectrophotometrically at 410 nm. Ortho-nitrophenyl-β-D-galactopyranoiside is used as a substrate; it gives a yellow coloured product (ortho nitrophenol) in the presence of β-galactosidase.

luciferase and β-glucuronidas

Luciferase and β-glucuronidase are two recently introduced reporter genes that are finding widespread use in plant systems.

The activity of luciferase on its substrate luciferin causes it to luminesce, the intensity of luminescence can be assayed in a luminometer.

β-glucuronidase activity can be assayed spectrophotometrically, fluorimetrically or histochemically. This type of assay is becoming very popular amongst plant molecular biologists because most plant species have only very low endogenous levels of activity of these enzymes.

Now that we have looked at which reporter genes are available, we can examine what they can be used for. We will look at one 'real life' example to illustrate the use of reporter genes. The example we will look at is the study of cell specific expression of insulin and chymotrypsin. The workers involved wanted to know if the promoter and 5' flanking regions of the genes were responsible for the cell specific expression. Genes encoding insulin and chymotrypsin had already been cloned, so the 5' flanking regions and promoter regions of each gene could be joined to the chloramphenicol acetyl transferase coding region which acted as a reporter gene. These gene constructs were then used to transform cultured cells. The cells used were pancreatic endocrine and exocrine cells. *In vivo* insulin is synthesised only in endocrine cells and chymotrypsin only in exocrine cells. When the activity of chloramphenicol acetyl transferase was assayed, it was found that the 5' flanking regions of the insulin gene only conferred expression (that is chloramphenicol acetyl transferase activity) in cultured endocrine cells. The 5' flanking regions of the chymotrypsin gene only gave expression in exocrine cells. Deletion mutagenesis of the 5' flanking regions showed that this cell specific expression was due to a short piece of DNA located between 150 and 300 basepairs upstream of the transcription start site.

investigation of cell-sepcific expression of insulin and chymotrypsin genes

SAQ 8.3

You have what you know to be a partial genomic clone of a gene which you suspect is expressed in a tissue specific manner. The clone does not contain any promoter sequences nor the complete coding region.

Listed below are several steps involved in a simplified scheme to isolate a full length clone and study promoter function. There is a question associated with each step. List the steps in order and answer the associated question. Many of the questions do not have simple right/wrong answers so think carefully.

1) Transform an embryo of an organism and use a histochemical reaction to check if the promoter fragments give tissue specific expression in the mature organisms.

   What reporter gene would be particularly useful for use in plants?

2) Restriction map the clone containing the promoter regions and identify fragments suitable for subcloning into a vector containing a reporter gene.

   What features might such a fragment have to ease subsequent manipulations?

3) Mutagenise the fragment *in vitro* to identify a small region responsible for the tissue specific expression.

   What mutagenesis techniques might you use?

4) Initiate chromosome walking to isolate overlapping clones.

   What type of sequences may cause walks to split?

5) Identify clones containing the complete coding sequence and promoter fragments. Use nuclease protection assays and primer extension to identify RNA termini.

   What is the advantage of primer extension?

8.5 Protein/DNA interactions

It is known that many proteins are involved in regulating gene expression. Very often such DNA/protein interactions occur in the promoter region or in other gene-flanking regions. We will, therefore, turn our attention to DNA/protein interactions in the final section of this chapter. We will look at several methods for investigating protein/DNA interactions, including filter binding assays, gel retardation assays, nuclease protection assays and 'footprinting'.

We will begin by looking briefly at how potential DNA binding proteins are isolated from cells. Nuclei are usually used as the source of such proteins. The nuclei preparation used can sometimes be simply a crude extract or more usually consists of nuclei that have been purified by percoll gradient centrifugation.

extraction of proteins from nuclei

Proteins are isolated from nuclei in a high salt buffer that contains inhibitors of proteases that are often present in nuclei preparations. The extract is then centrifuged and the supernatant which contains the nuclear proteins is dialysed so as to reduce the salt concentration. The protein concentration can then be determined.

8.5.1 Filter binding assays

Filter binding assays are based on the observation that, in the presence of magnesium ions, double-stranded DNA usually fails to bind to nitrocellulose filters, whereas proteins tend to be bound to the filter. If a piece of DNA carries a protein that is bound to it, it will also be retained on the filter.

The DNA used in filter binding assays is usually derived from recombinant cosmid or λ phage vectors by restriction digestion. A mixture of differently sized DNA fragments can be used as later steps in the procedure allow separation and visualisation of all the DNA fragments retained on the filter. To allow this visualisation, the DNA fragments used are usually labelled on both 3' and 5' ends with ^{32}P.

8.5.2 Gel retardation assays

separation of complexed and uncomplexed DNA

Gel retardation assays are based on the fact that electrophoresis through low ionic strength gels (which can be either agarose or polyacrylamide) allows protein-DNA complexes to be resolved from uncomplexed DNA. It has the major advantage over the filter binding assay that it can also detect the interactions of several different proteins that bind to a single piece of DNA. This procedure is also about ten-fold more sensitive than the filter binding assay. One drawback of gel retardation assays, however, is that only relatively small pieces of DNA can be used as probes (up to about 300 basepairs), whereas in the filter binding assay pieces of DNA larger than 1000 basepairs can be used. This may effectively mean that gel retardation assays can only be performed on regions of DNA that are already characterised in some way so as to allow a more 'targeted' analysis.

The DNA used in gel retardation assays is usually only labelled at one end. In the initial stages the technique is more-or-less identical to the filter binding assay, with labelled DNA being incubated with a preparation of nuclear proteins. This reaction mixture is then applied to a gel of either agarose or polyacrylamide and subjected to electrophoresis. The labelled DNA can then be visualised by autoradiography. The migration of the uncomplexed probe can subsequently be compared with the migration of protein - DNA complexes.

8.5.3 *In vitro* 'footprinting'

DNA protected
from DNase I
and
exonuclease III
by bound
protein

In footprinting, two nucleases, exonuclease III and DNase I, are used for identifying the sites where proteins are bound to DNA. The basic method used depends on the fact that proteins bound to a piece of DNA can protect it from digestion by the nuclease which is being used. Although both procedures will seem to be relatively simple it is often necessary to characterise the amount of protein binding and optimise conditions quite carefully before commencing either exonuclease III or DNase I footprinting. This technique is called footprinting because it is as though the bound protein is left as 'footprints' on the DNA.

We can represent this in the following way:

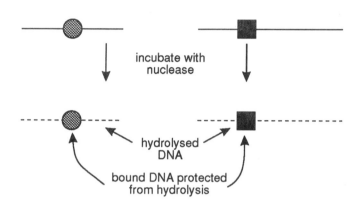

We also call this technique, nuclease protection assays. It has many similarities to the protection of DNA by complementary RNA as described in Section 8.2.2.

We can look at a procedure for DNase I digestion to illustrate how nuclease protection footprinting works.

End-labelled fragments of DNA are incubated under clearly defined conditions with extracts of nuclear proteins. DNase I is then added and the incubation continued such that only partial digestion occurs. These partial digests can be separated by gel electrophoresis and protein DNA complexes identified by autoradiography. The gel bands containing the protein-DNA complexes can be cut out. The DNA can also be fractionated by denaturing gel electrophoresis. The pattern produced by the DNase I digests can then be compared with the pattern produced by digesting the same DNA in the absence of nuclear proteins. From this, we can identify DNA fragments which are protected from DNase I digestion (by bound proteins).

Finally we will look at a technique known as *in vivo* footprinting.

8.5.4 *In vivo* footprinting

In vivo footprinting is a recent technique that can be used to accurately map the position of protein DNA complexes *in vivo*.

Basically the technique depends on cleaving genomic DNA (in intact nuclei or in whole cells) with either nucleases such as DNase I or with chemicals. In areas where a protein is bound to the chromosome, the DNA will be protected from cleavage, thus the protein will 'leave a footprint'. The partially digested DNA produced is then cut with restriction endonucleases and hybridised to a probe that is complementary to the genomic DNA. The probe DNA used is usually obtained from a clone of the gene of interest, and thus its position can be determined relative to (for instance) the start of transcription. After hybridisation and purification, the hybrids of the latter are electrophoresed and visualised by autoradiography. The pattern of cleavage products produced by treatment with the nuclease or the chemicals *in vivo* can be compared with the pattern obtained by digestion of extracted protein-free genomic DNA.

purification of
DNA-binding
proteins

The methods we have looked at here allow the sites of protein - DNA complexes to be mapped, and as well as allowing identification of protein binding regions in DNA, may permit purification of the protein that is bound to the DNA. Techniques such as these can prove extremely useful in elucidating mechanisms controlling gene expression. It may be possible for instance to isolate sequences responsible for tissue specific gene expression and identify, isolate and characterise any proteins binding to that particular region of the promoter. These proteins may be found to control the expression of the gene in some way. The opportunity for engineering specific promoters for specific tasks is thus coming that much closer.

SAQ 8.4

For each of the pairs of statements below indicate from the list which technique is applicable to each pair of statements:

1) can be used with large fragments of DNA;

   cannot distinguish how many proteins are bound to a DNA fragment;

2) can only be used with small DNA fragments;

   allows purification of the protein;

3) gives accurate indications of protein DNA complexes *in vivo*;

   requires comparison with deproteinised DNA;

4) compares the cleavage produced with and without added nuclear proteins *in vitro*;

   allows purification of the protein;

   a)   *in vivo* footprinting;

   b)   filter binding assay;

   c)   gel retardation;

   d)   *in vitro* footprinting.

Summary and objectives

In this chapter, we first described how clones can be isolated and how a gene can be localised within cloned DNA. We then dealt with methods that can be used to locate the termini of genomic sequences and then described the chromosome walking techniques for isolating overlapping DNA sequences from a library. In the final sections, the techniques used in studying regulatory regions and protein/DNA interactions were given.

After reading this chapter, you should be able to:

- describe how genomic clones can be isolated and how coding regions can be identified;

- select techniques to locate gene termini;

- describe how full length clones can be isolated by chromosome walking;

- describe how regulatory sequences can be identified and characterised, including the use of reporter genes;

- select appropriate reporter genes;

- describe how protein/DNA interactions can be studied;

- select appropriate techniques to identify sites of protein - DNA interactions.

Naturally-occurring organisms, categorisation and containment

Naturally-occurring organisms, categorisation and containment

9.1 Introduction

The first eight chapters of this text have been devoted to describing the techniques used to manipulate genetic material to produce new genes or new arrangements of genes. For the most part, such manipulations have involved the use of micro-organisms. Of prime concern to those involved in the process of genetic manipulation and to the community at large are the issues of safety. It is undeniable that such procedures can potentially present a great variety of hazards. To take an extreme example, if the gene coding for a botulin toxin was introduced into a bacterium which was a normal member of the human gut flora and this was released into the environment, the results could be catastrophic. Humans are not the only ones at risk: the generation of plant or animal pathogens of greater virulence and a plant population with a greater tendency for weediness may also have disastrous consequences for the biosphere. It is, therefore, entirely appropriate for us to consider the safety issues of genetic manipulation in some detail. Here we have divided the discussion into two parts.

In this chapter, we discuss the safety issues associated with the use of micro-organisms. This reflects both the importance of micro-organisms in biotechnology and their importance as potential mediators of diseases of mankind and other biological entities. In the following chapter, we will discuss the safety issues concerned with genetically manipulating organisms and the assessment of the risks involved in constructing organisms displaying changed characteristics.

In order to be able to understand the biosafety issues involved in microbiological biotechnology and genetic engineering we must consider the following points:

- the mechanisms by which the human body may be harmed by pathogenic bacteria and the means by which it may defend itself;

- aspects of safety regulations and law relevant to this discussion, including the categorisation of pathogens according to hazards and categories of containment that are required to prevent the escape of organisms;

- the use of biological safety cabinets;

- methods of disinfection and sterilisation.

international directives, recommend-ations and agreements

national regulations

Before we embark on this discussion, you should realise that the standards that are applied in these areas of activities are the subject of a range of international directives, recommendations and agreements which manifest themselves in the form of a variety of national regulations. These regulations are in a state of evolution. Despite such changes and the differences between individual national regulations, there is still a considerable amount of common ground about the safety conditions that are applicable in this area. Bearing this in mind, we provide guidance to these generally accepted conditions and support this by some specific examples. This avoids the necessity of providing a tedious, encyclopedic approach but ensures that you have the necessary knowledge and develop an appropriate awareness and attitude to use micro-organisms and genetic manipulation safely. It is, of course, incumbent on all workers to conform to all of the relevant regulations in operation in the region in which the work is being undertaken.

9.2 Natural organisms in the environment

There are, in fact, only a few sites on the surface of the earth where micro-organisms do not exist. The majority of objects, which have not been decontaminated by some sterilisation process, harbour a multitude of different micro-organisms.

niche

In nature, many organisms live in specific places and fulfil a particular role in the environment. Such a specific place is known as a niche. Large numbers of organisms, such as moulds and bacteria, multiply side by side in the environment until a characteristic population or associative flora has developed. Many different niches exist. In the soil, for instance, growing micro-organisms are found even at low temperatures and low concentrations of water-soluble nutrients. Microbial life is widespread and comprises many thousands of species which live independently by metabolising dead organic material. Such organisms are called saprophytes. However other species have adapted to life with animals, plants or Man. They either live in symbiosis with the host as useful commensals (an organism, usually a bacterium that benefits by living within or on the surface of another organism, usually an animal, without causing disease) or are pathogens which can cause disease (see Section 9.5).

saphrophytes

commensals

pathogens

Saprophytes and commensals are generally harmless to Man. The pathogens can be classified into groups of different risk to Man, dependent on the gravity of the disease they induce and their ease of spread.

SAQ 9.1

Define the following terms by writing the number 1, 2 or 3, representing one of the three definitions, in the boxes provided.

pathogen ☐

saprophyte ☐

commensal ☐

1) organism which metabolises dead organic matter;

2) organism which benefits by living within or on the surface of another organism without causing disease;

3) organism which causes disease.

9.3 Historical use of natural organisms

Throughout history, Man has used certain harmless organisms in the production of food, for example in the production of beer, yoghurt and cheese. The large-scale production of such foodstuffs is nowadays commonplace. High levels of hygiene and cleanliness are required so as to avoid the introduction of unwanted contaminating micro-organisms into the process. When using 'safe' organisms, little is done to prevent, in an absolute sense, their escape into the environment.

9.4 Natural organisms with associated risks

Of course not all industrial or pharmaceutical processes involve such 'user friendly' organisms and some of the micro-organisms present in the environment are, as we said, pathogenic. We must be aware that the use of pathogenic organisms, unless under stringently controlled conditions, would be risky in the extreme. It is of course vitally important that people considering this type of undertaking are able to assess the risks and balance them against the potential benefits. To this end, many countries have strictly enforced laws and regulations. In order to comply with legal/safety requirements, people engaged in such undertakings must be in a position where they can understand the potential risks involved and appreciate the possible dangers and the type of precautions that are required to eliminate, or minimise these risks.

In the following section we will discuss some of the factors involved in the pathogenicity and virulence of an organism and the manner in which it is dangerous to its host.

Later on we will relate the need for biological safety in the use of natural organisms to their classification into four hazard groupings, associated precautions and procedures for their safe use.

9.5 Infectious diseases, pathogenicity and virulence.

9.5.1 The body's defence mechanisms

In this section, we are going to summarise some background information regarding the events leading from the contact of a host with a pathogenic micro-organism to infection and illness. A fuller description of defence against such infections is given in the BIOTOL text, 'Defence Mechanisms'. Details of the immune system are provided in the BIOTOL text, 'Cellular Interactions and Immunobiology'.

the skin is an effective barrier

All infectious diseases start at one of the surfaces of the host, the skin, the gastrointestinal tract or the respiratory tract. Intact skin is a remarkably effective barrier to infection. Spraying of bacterial suspensions on to the skin usually has no deleterious effect on the skin because the concentration of acids and salt and the low pH of the healthy skin kills a large variety of bacteria. The resident flora of the skin which is adapted to these conditions may also produce bactericidal substances. But cuts and abrasions will locally remove the physical barrier of the skin. Many organisms can then enter and cause local inflammation. The larger the wound, the greater the chance of deeper and general infections.

protective action of acid in the stomach

The gastrointestinal tract is protected to a large extent by the hydrochloric acid in the stomach which has bactericidal properties although it may be weakened by the ingestion of food with neutralising properties. An important property of the gastrointestinal tract is that food is transported rapidly through it, which means that potentially risky bacteria are rapidly expelled and are not in residence for long. The gastrointestinal tract is coated with a mucosal layer which contains selected antibodies which also provide defence against infection. In addition, the lower parts of the tract have their own immense and specific bacterial flora, which gives little chance for development of most pathogens.

importance of commensal organisms of the gastrointestinal tract

protection by mucous and phagocytic cells in lungs

The respiratory surface has its own physical protection. It produces a mucous coating that traps most types of microbial life. The entrapped cells can either be expelled with the mucous or be killed by phagocytic cells. These cells form protrusions called pseudopodia which fuse around the pathogen to form a phagocytic vesicle or phagosome in which the invader is killed. The mucous coating is continuously being swept up the respiratory passages by cilia, and it is coughed up and subsequently swallowed, thereby depositing potentially infectious material into the bactericidal contents of the stomach.

internal defences include cells and humoral factors

If microbes breach the body's physical defence system, they encounter an extremely hostile environment in which humoral factors (antibodies) and phagocytic and other white blood cells prevent bacterial growth. Only bacteria which are resistant to these defences are capable of causing inflammation or disease.

pathogenicity

There are only a few hundred strains of pathogens for Man able to incite disease in susceptible hosts. Pathogenicity for a defined host is a qualitative property: a micro-organism is either pathogenic or not (see Figure 9.1).

Figure 9.1 A guinea pig and a rat have been infected with equal doses of diptheria bacilli. The guinea pig is very ill, the rat is not susceptible. The diphtheria bacilli are not pathogenic to rats.

virulence

Virulence is measured by the number of organisms required to damage or kill the host. The virulence of a strain of a pathogen is a quantitative property: strong or weak. Virulence is dependent on the infective dose, on the growth rate of the pathogen in the body and on the degree of toxicity. The infective dose of a few micro-organisms is very low, for example those which cause respiratory infections often have especially low

ID50

ID50 values. ID50 is the infective dose which will cause infection in 50% of the individuals tested.

Table 9.1 shows the comparative ID50 values for a few micro-organisms which cause respiratory tract infections compared to examples of other organisms which have relatively high ID50 values.

---

a) ID50 values for micro-organisms which cause respiratory tract infections

Organism	ID50 value
adenovirus	1 virion* (for humans only)
influenza virus	3 virions
coxsackie virus	28 virions
Mycobacterium tuberculosis	1-10 bacteria

b) ID50 values for micro-organisms which may cause gastrointestinal tract infection

Organism	ID50 value
Shigella	200 bacteria
Salmonella	107 bacteria

* A virus consists of nucleic acid and protein. The complete virus is named a virion, a term which denotes the intactness of structure and the property of infectiveness.

Table 9.1 Examples of ID50 values for a variety of respiratory and gastrointestinal tract infections.

SAQ 9.2

Write down the answers to the following questions concerning the body's defence mechanisms.

1) Which antibacterial factors are present on human skin?

2) What are the defence mechanisms present in the gastrointestinal tract?

3) What is the defence mechanism of the respiratory tract?

4) What function do phagocytic cells perform?

9.5.2 Specific properties of pathogens

A pathogen is able to survive and multiply on or in the surfaces or tissues of the host. Growth of a pathogen in the body requires special properties for attachment to the host cell, penetration into the cell, growth under the conditions found within the tissues (temperature, pH, levels of oxygen, nutrient concentration), evading or resisting defence mechanisms and the production of toxins. Many genes are required in the infecting organism for it to be capable of survival and growth.

A pathogen should be able to pass the surface barrier and enter the body's tissues. Growth does not always imply that the pathogen spreads through the body. Diphtheria bacteria remain in the throat whilst the exotoxin spreads. In general, bacterial spread is limited by the mucous membranes and phagocytosis. Some pathogens evade phagocytosis, they may even be able to survive in phagocytic cells and spread along the

evasiveness   lymph and blood vessels. This is called evasiveness.

In contagious diseases, the pathogen has to get outside the body to infect new hosts (for example influenza by coughing and sneezing). Sometimes the pathogen is excreted in the faeces or wound secretions.

9.6 Bacterial toxins

exotoxins

Pathogens produce a variety of extracellular metabolites, enzymes and toxic proteins which can interfere with host defence mechanisms or damage host cells. The alpha-toxin of *Staphylococcus aureus* which causes haemolysis (lysis of red blood cells) is a specific example. A few pathogens grow locally and produce extremely toxic exotoxins, which spread through the body causing the symptoms of the disease, examples are tetanus and diphtheria. Immunisation against such diseases may be possible and is discussed in Section 9.8.

endotoxins

Some toxic products of bacteria are associated with the bacterial cells and are not excreted by these cells. Endotoxins are produced by Gram negative bacteria. They consist of antigenic complexes of polysaccharides and lipids derived from the cell wall. These toxins are more heat stable but less toxic and cause less specific symptoms (for example, fever or nausea).

9.7 Transmission of human pathogens

obligate and facultative parasites

Human pathogens generally do not multiply outside their host. They are mostly obligate parasites with patients or carriers acting as the source of infection. Only a few wound infections are caused by free living bacteria, for example tetanus. We may call these bacteria facultative parasites.

Enteric infections may be spread by faecal matter. In some foods such pathogens can multiply to well over their ID50 values, which may cause food infection, often inaccurately called food poisoning. The distinction between food infection and food poisoning is as follows:

- food infection is the contagious illness caused by foods in which a viable pathogen is present. Examples are salmonellosis (*Salmonella*) and bacillary dysentery (*Shigella*). Some enteric viruses can also be transmitted with food, though they do not multiply in food (hepatitis A, poliomyelitis);

- food poisoning is the illness caused by ingestion of foods containing bacterial and other toxins and is generally not contagious. Such toxins are produced for example by *Staphylococcus aureus* or *Clostridium botulinum*.

Respiratory infections including influenza, scarlet fever, whooping cough and diphtheria are spread by aerosols or by contact.

Examples of infections transmitted by insects are malaria and yellow fever. Hepatitis B and AIDS may be spread by blood-contaminated syringes.

formites

Wounds may become infected externally by the wound-causing object itself or by pathogens present on clothing. Formite is a general term which is often used to describe items such as bedding, clothing and utensils which may carry infectious organisms. Wounds are very often infected by *Staphylococcus aureus* from the patient's own nose or skin. *Staphylococcus aureus* strains cause many skin infections, but they may also invade and produce severe infections in any other tissue or organ of the body. In their adaptation to parasitism, they have been among the most versatile and, notwithstanding the application of numerous anti-staphylococcal antibiotics in the past 35 years, they remain a serious problem. Most of the serious staphylococcal infections currently encountered are seen in patients whose normal defences are severely impaired.

SAQ 9.3

Write down your answers to the following:

1) How are the pathogens causing influenza spread?

2) What is a virion?

3) What is microbiological food poisoning?

4) What is food infection?

5) Can the following produce bacterial toxins? Ring where appropriate.

Clostridium botulinum	Yes/No
Staphylococcus aureus	Yes/No
The organism which causes diphtheria	Yes/No
The organism which causes tetanus	Yes/No

9.8 Prevention of disease

Children are susceptible to, and catch, large numbers of diseases. The body usually overcomes these infections and develops immunity to them so that by maturity most adults are immune to natural infections that are common childhood ailments.

immunisation

toxoids

Immunisation against certain diseases is possible. The type of immunisation varies depending on the disease. Treatments to prevent diseases such as tetanus and diphtheria involve the use of inactivated exotoxins, known as toxoids, prepared from the exotoxins by treatment with formaldehyde or heat. For safety reasons, the diphtheria strain used for the production of diphtheria toxin has been treated, causing it to lose the properties that enable it to grow in the body.

As we discussed before, endotoxins are more stable and less specific in their action. They do not form toxoids and cannot be used for immunisation.

prevention of spread

Many diseases which are spread by ingestion of contaminated food can be prevented from spreading by blocking the routes of transmission such as by the use of sewage systems, water purification, pasteurisation of milk and kitchen hygiene.

In cases where the disease is spread by an insect vector, the presence or absence of the vector in a particular location is obviously an important factor. In some areas, the control of fresh outbreaks of malaria has been achieved by destroying the breeding grounds of the mosquito.

You should now be aware of the many risks posed to laboratory workers and the community by the careless use or misuse of micro-organisms. It is for this reason that the law insists that all risks are minimised, wherever possible, through the use of appropriate containment facilities and vaccination of workers where necessary.

9.9 Classification of micro-organisms into risk (hazard) groups

Organisms may be categorised into groups depending upon the hazards they pose to the laboratory worker and the community in general. In this chapter we will deal with naturally occurring organisms and in Chapter 10 we will consider genetically manipulated organisms.

Micro-organisms may be categorised into hazard groups by using a framework of criteria such as:

- is the organism pathogenic for Man?

- is it a hazard to laboratory workers?

- is it transmissible to the community?

- is effective prophylaxis or treatment available?

These hazard groups are formed according to the infective hazard that the organisms within the group pose for healthy workers. It must be emphasised at this stage that the groupings do not take into account any additional risk that an organism may present for those persons with pre-existing disease, compromised immunity, those who are pregnant or effected by medication, nor does it take account of the allergenic properties of the organism or of toxic hazards from its products.

The European Federation of Biotechnology (EFB) in 1989 introduced four EFB risk classes of naturally occurring micro-organisms. These are:

- harmless micro-organisms;

- low-risk micro-organisms;

- medium-risk micro-organisms;

- high-risk micro-organisms.

A description of these categories is provided in Table 9.2. Read through this table carefully.

Harmless micro-organisms have an extended history of safe use in human consumption (eg beer, yoghurt, cheese) and on a large scale in the biotechnology industry (antibiotics, enzymes, amino acids) as well as in agricultural applications (*Bacillus thuringiensis*).

EFB Class 1 is defined as:

'Those micro-organisms that never have been identified as causative agents of disease in Man and that offer no threat to the environment. They are not listed in higher classes'.

Low-risk micro-organisms do not easily survive outside the host, are unlikely to spread and do not cause serious diseases.

EFB Class 2 is defined as:

'Those micro-organisms that may cause disease in Man and which might, therefore, offer a hazard to laboratory workers.

They are unlikely to spread in the environment.

Prophylactics are available and treatment is effective.'

Medium- and **high-risk** micro-organisms are severe pathogens and they are rarely used for large-scale operations. On a small scale, they are used for the production of vaccines or diagnostics.

EFB Class 3 is defined as:

'Those micro-organisms that offer a severe threat to the health of laboratory workers but a comparatively small risk to the population at large.

Prophylactics are available and treatment is effective.'

EFB Class 4 is defined as:

'Those micro-organisms that cause severe illness in Man and offer a serious hazard to laboratory workers and to people at large. In general effective prophylactics are not available and no effective treatment is known'.

Table 9.2 EFB risk classes of naturally-occurring micro-organisms.

∏ Why are the low-risk pathogens (EFB 2) not dangerous if their release is kept at the low (= minimal) level found in laboratories when good microbiological practices are adhered to?

You should have realised that good microbiological practices strictly limit the release (or escape) of micro-organisms. Thus the number of micro-organisms likely to infect a laboratory worker are much lower than the ID50 concerned. Thus, the individual is unlikely to develop the disease.

The four EFB risk classes are reflected in many national-based groupings. Here we will give just two examples. The UK Advisory Committee on Dangerous Pathogens (ACDP) is an agency of the UK Health and Safety Executive. They define four hazard groups (see Table 9.3) which have great similarity to those defined by EFB. Similarly in The Netherlands, four groups are defined (Pathogen Groups 1,2,3,4 - often abbreviated to PG1,2,3,4).

Table 9.3 shows ACDP hazard groups. Compare these groupings with those given in Table 9.2.

Hazard Group 1

An organism that is most unlikely to cause human disease.

Hazard Group 2

An organism which may cause human disease and which might be a hazard to laboratory workers but is unlikely to spread to the community.

Laboratory exposure rarely produces infection and effective prophylaxis or effective treatment is usually available.

Hazard Group 3

An organism that may cause serious human disease and presents a serious hazard to laboratory workers. It may present a risk of spread to the community but there is usually effective prophylaxis or treatment available.

Hazard Group 4

An organism that causes severe human disease and is a serious hazard to laboratory workers. It may present a high-risk of spread to the community and there is usually no effective prophylaxis or treatment.

Table 9.3 ACDP hazard groups.

SAQ 9.4

Which of the EFB risk classes and ACDP hazard categories of organisms may be referred to as pathogens?

The lists of organisms in the risk classes (hazard groups) are very long and cover: bacteria, chlamydias, rickettsias, mycoplasmas, fungi, parasites and viruses. Certain strains of organisms which are listed may justify different levels of containment than the hazard ratings suggest. This applies to mutants of increased or decreased virulence, attenuated vaccine viruses or bacteria and antibiotic resistant strains (it also applies to genetically manipulated bacteria - which we will deal with in Chapter 10). In cases where many organisms have been grouped together certain members (species or strains) may justify the use of different levels of containment than is necessary for the group as a whole.

We have provided a few examples of specific organisms from some of the Hazard Groups in Appendix 3. This list is based on the recommendations of the ACDP in the UK. There is, however, more-or-less international agreement on the hazards (risks) associated with the named examples. You should, of course, become acquainted with the specific details of the recommendations in your region. Particular states impose specific conditions on some pathogens. For example, deliberate cultivation of the *Variola* (smallpox) virus is banned totally in some countries. The reader should not assume that the absence of a particular species from our list means that the organism is free from risk. A key aspect to working with micro-organisms is the need to be aware,

different restrictions may apply in different states

not only of the biological hazards but also of the regulatory obligations that must be fulfilled.

plant pathogens It is not only human pathogens that are covered by such regulations. Similar restrictions apply to work with plant pathogens. These too are categorised, although the risk of plant pathogens depends more on local situations, for example is the plant pathogen endemic in the area or is it exotic? High-risk plant pathogens are quarantine organisms. In the UK for example, it is the responsibility of the user to determine whether or not the organism they wish to use is a plant pathogen or pest, whether or not it is indigenous, and to apply for a licence to use the organism if appropriate. Where possible the user will be helped to select a lower risk organism.

The EFB has published for plant pathogens, a risk classification (Küenz *et al* 1987, Appl Microbial Biotechnol 27,105).

There are three classes: low-, medium- and high-risk. For low- and medium-risk pathogens, the Plant Protection Authority should always be consulted. High-risk plant pathogens are quarantine organisms; they should not be used (see Table 9.4).

Ep1	low-risk if not endemic
Ep2	medium-risk for crops
Ep3	high-risk quarantine organisms

Table 9.4 EFB classes of plant pathogens.

Similarly, the use and importation of any pathogen that may cause disease in agricultural animals, birds, fish and bees is strictly controlled.

Regulatory aspects of the importation/use of these pathogens vary from county to country and are the subject of continuous updating.

9.10 Safety precautions

In the remainder of this chapter we will concentrate on the requirements for work at various levels of containment.

At the entrance to laboratories carrying out work with biologically hazardous material the sign shown in Figure 9.2 should be shown.

Figure 9.2 International Biohazard sign.

9.11 Classes of containment

In the previous sections, we have explained that micro-organisms can be divided into four risk classes (hazard groups). The extent of containment that needs to be maintained reflects the risks associated with each of these groups. As a general principle, the chance of escape should be **minimised** for low-risk micro-organisms and should be **prevented** for medium and high-risk micro-organisms. These safety precautions can be divided into three aspects:

* procedures;

* primary containment (equipment);

* secondary containment (facilities).

harmonisation of national regulation

Again there are national differences in the details of the recommendations and regulations which are applied to these issues although they have many features in common. International bodies such as European Federation of Biotechnology (EFB) and the Organisation for Economic Co-operation and Development (OECD) are doing much to harmonise the different national risk classifications and containment systems. Here we will take a generally accepted position. It is, however, important that readers are familiar with their own national situation. For example, the current Netherlands Recommendations for Safe Microbiological Work are published in Aanbevelingen voor Veilig Microbiologisch Werk 2nd ed. Editors: Nederlandse Vereniging voor Microbiologie, RIVM Bilthoven. In the UK these are published by HMSO (Her Majesty's Stationery Office) in a publication entitled, 'Advisory Committee on Dangerous Pathogens - Categorisation of Pathogens According to Hazard and Categories of Containment 1990'.

9.11.1 Harmless micro-organisms

For the use of harmless micro-organisms (EFB1, Hazard Group 1, PG 1), there exist only limited containment requirements.

Such organisms can, for example, be used in industrial processes such as food manufacture. In such circumstances, general, good microbiological practices are usually sufficient. These are required to protect cultures from contamination and to provide quality assurance. Similarly in the laboratory, basic microbiological techniques (BMT) are sufficient to ensure the absence of contaminants (use of sterile culture media, aseptic inoculation). We have provided a model set of laboratory rules for the containment of micro-organisms in this category in Table 9.5. We will call this Containment Level 1 (consistent with ACDP nomenclature).

BMT

When handling this group of organisms on an industrial scale, for example in the manufacture of such foods as yoghurt, cheese, mycoprotein and food enzymes, analogous standards are applied. A set of measures called Good Industrial Large-Scale Practices (GILSP) have been devised by the OECD.

GILSP

___ The laboratory should be easy to clean. Bench surfaces should be impervious to water and resistant to acids, alkalis, solvents and disinfectants.

___ If the laboratory is mechanically ventilated, it is preferable to maintain an inward airflow into the laboratory by extracting room air to atmosphere.

___ The laboratory must contain a wash-basin or sink that can be used for hand washing.

___ The laboratory door should be closed when work is in progress.

___ Laboratory coats or gowns should be worn in the laboratory and removed when leaving the laboratory suite.

___ Eating, chewing, drinking, smoking, storing of food and applying cosmetics must not take place in the laboratory.

___ Mouth pipetting must not take place.

___ Hands must be disinfected or washed immediately when contamination is suspected, after handling viable materials and also before leaving the laboratory.

___ All procedures must be performed so as to minimise the production of aerosols.

___ Effective disinfectants must be available for immediate use in the event of spillage.

___ Bench tops should be cleaned after use.

___ Used laboratory glassware and other materials awaiting disinfection must be stored in a safe manner. Pipettes, if placed in disinfectant, must be totally immersed.

___ All waste material which is not to be incinerated should be rendered non-viable before disposal.

___ Materials for disposal must be transported in robust containers without spillage.

___ All accidents and incidents must be recorded.

Table 9.5 Model laboratory rules for Containment Level 1. These rules are typical of those which are applied to handling micro-organisms classified as being harmless (eg EFB1, UK-Hazard Group 1, Netherlands-PG1). These rules are based on those recommended by the Advisory Committee on Dangerous Pathogens in the UK and are designed to minimise the **release** of micro-organisms during laboratory work.

9.11.2 Low-risk micro-organisms (EFB2, Hazard Group 2, PG2).

GMT

Good Microbiological Techniques (GMT) are essential in all work with low and higher risk micro-organisms in the laboratory. In contrast with BMT, which is mainly concerned with ensuring that cultures are transferred without contamination, GMT provides the laboratory worker and the environment against contamination and possible infection from cultured pathogens.

Many of the features shown in Table 9.5 are also operational at this level. There are, however, some important additions. For example, in the UK it is recommended that $24m^3$ of air space is available for each worker, wash basin taps must be of a type that can be operated without being touched by hand and bench tops must be disinfected after use. It is also demanded that for manipulations such as vigorous shaking, ultrasonic disruption and other techniques that create aerosols, a microbiological safety cabinet or equipment designed to contain the aerosol must be used. There should also be restricted access to the facility.

In the UK, this level of containment is called Containment Level 2. EFB categorisation regards this level as Containment Category 1 (CC1). The equivalent in The Netherlands is Fysisch Inperkings Nivo 1 (FIN1).

9.11.3 Medium-risk micro-organisms (EFB3, Hazard Group 3, PG3)

This hazard group is regarded as Containment Category 2 (CC2) by EFB, as FIN2 in The Netherlands and Containment Level 3 in the UK. The conditions imposed under these categorisations are, however, very similar. We have highlighted the major measures that have to be taken in addition to those taken at lower containment levels in Table 9.6.

— only authorised personnel are admitted to the facility

— if vaccines are available, personnel are vaccinated

— if air is extracted from the facility, HEPA filters are used

— effluents from the facility should be decontaminated or sterilised

— an autoclave should be within the facility

— all processes involving medium-risk micro-organisms must be carried out in hermetically-sealed equipment or biosafety cabinets

— protective suits, closing at the back, have to be worn by personnel

— hands and forearms should be washed and disinfected at regular intervals. No jewellery must be worn

---

Table 9.6 Measures taken in using medium-risk micro-organisms in addition to those taken with lower risk organisms. Note that with this category, the objective of the containment procedures is to **prevent** the release of micro-organisms during laboratory work.

9.11.4 High-risk micro-organisms (EFB4, Hazard Group 4, PG4)

Clearly use of organisms in this category must be done in such a way as to prevent any release. This implies prevention of release at **any** stage of procedures. Thus equipment and facilities must be designed to prevent any escape (ie full primary and secondary containment devices must be used). Only experienced personnel are allowed to handle such organisms and they must receive extensive training. Their work must be supervised. In Table 9.7 we highlight some features of this level of containment.

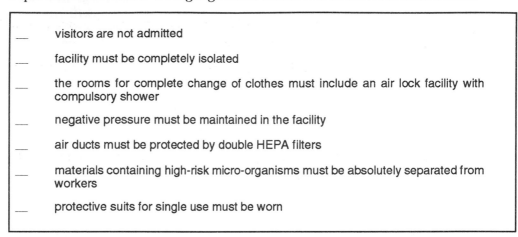

	visitors are not admitted
__	facility must be completely isolated
__	the rooms for complete change of clothes must include an air lock facility with compulsory shower
__	negative pressure must be maintained in the facility
__	air ducts must be protected by double HEPA filters
__	materials containing high-risk micro-organisms must be absolutely separated from workers
__	protective suits for single use must be worn

Table 9.7 Measures taken in using high-risk micro-organisms in addition to those taken with lower risk organisms. Note that under EFB nomenclature this level of containment is described as Containment Category 3 (CC3). In The Netherlands it is described as FIN3 whilst in the UK it is referred to as Containment Level 4. The aim of these measures is to be absolutely certain that organisms are **prevented** from escaping from the laboratory.

The following chart (Table 9.8) is a summary of the laboratory containment requirements for harmless (EFB Class 1), low-risk (EFB Class 2), medium-risk (EFB Class 3) and high-risk (EFB Class 4) micro-organisms. Y represents yes, meaning that the requirement is needed, N is for not required.

Containment requirements	EFB Class of micro-organism			
	1	2	3	4
laboratory site:isolation	N	N	partial	Y
laboratory: sealable for fumigation	N	N	Y	Y
airlock	N	N	optional	Y
airlock with shower	N	N	N	Y
wash basin	Y	Y	Y	Y
effluent treatment	N	N	optional	Y
autoclave: in the laboratory suite	N	Y	Y	Y
in double-ended laboratory	N	N	N	Y
microbiological safety cabinet/enclosure	N	optional	Y	Y

Table 9.8 Some features of containment requirements with micro-organisms of different risk classes.

SAQ 9.5

1) Which risk groups of micro-organisms should personnel be vaccinated against if suitable vaccines exist?

2) Which risk groups of micro-organisms must hermetically-sealed equipment or biosafety cabinets be used with?

3) With which risk groups of micro-organisms should the generation of aerosols be minimised?

4) With which risk groups of micro-organisms must all accidents and incidents be recorded?

5) With which risk groups of micro-organisms should a complete change of clothes be provided in an airlock facility?

SAQ 9.6

1) What hazard group is the organism *Staphylococcus aureus* in?

Ring the correct answer.

Hazard Group 1 2 3 4

2) What containment level should be used for work involving *Staphylococcus aureus*?

Ring the correct answer.

Containment Level 1 2 3 4

3) If a culture of *Staphylococcus aureus* grown up in the laboratory has to be vigorously mixed, what precautions must be taken and why?

4) If the bacterial mass is to be disposed of, what treatment is required?

5) What is the difference between contamination and infection?

9.12 Training

A key feature of all the various recommendations in operation is that they call for suitable training of laboratory personnel. This training obviously involves instruction concerning the details of operations and in most instances includes the use of 'biosafety manuals' which give detailed procedural descriptions of operations being undertaken in the laboratory. The extent of this training naturally increases with higher risk organisms. There is also a general requirement that an appropriate standard of supervision of the work is maintained.

biosafety manuals

Before we leave this aspect, we re-emphasise the point that it is essential that all workers are aware of the risk category of the micro-organisms they are handling and are familiar with both the appropriate national regulations and the local rules governing their use.

9.13 Measuring the safety of working conditions in laboratories and plants

9.13.1 Air contamination

RCS and CABS

The amount of airborne contamination can be determined using the Reuter Centrifugal Sampler (RCS) or the Casella Airborne Bacteria Sampler (CABS). The former samples 40 litres of air per minute, the latter has a capacity of 700 litres of air per minute. These instruments may be used to confirm that safe limits for air contamination are being met.

9.13.2 Surface contamination

Bacterial contamination of surfaces in the workplace may be assessed using contact plates or by taking swab samples.

contact method

In the contact method, RODAC (Replicate Organism Detecting and Counting) plates are filled with 'plate count agar' so that there is a slightly convex surface rising above the rim. The plates are applied to the surface to be tested (for example work-table or hand) and then incubated. Representative colonies are subcultured for identification.

swab method

In the swab method, swabs are made by wrapping and binding cotton wool around the ends of glass rods. Each swab is wetted, placed into a test tube and sterilised. A defined part of the surface to be investigated is rubbed firmly with a wetted swab. The swab is then shaken with nutrient broth. Part of the nutrient broth is mixed with molten, nutrient agar and allowed to gel. Incubation follows and again representative colonies are identified. Results obtained from both methods correlate quite well but neither measures 100% of the surface contamination.

aerosol generation

Using both surface and air monitoring methods, it has been shown that some processes such as pouring cultures into flasks, blowing out pipettes during work, subculturing, and open centrifugation caused severe air and surface contaminations. These could be avoided by carrying out such operations in a biosafety cabinet, which retains all aerosol droplets.

Human errors in the techniques used for handling micro-organisms may be discovered if monitoring is carried out consistently. Apart from the aerosol-creating procedures discussed another type of contamination which is sometimes found is that produced by steaming product lines without filtering the exhaust steam output, or cleaning equipment before thorough disinfection.

filter efficiencies

The type of filter chosen to retain aerosolised bacteria must meet with the appropriate standards in the country of use. Filters may be tested by challenging them with known levels of bacteria and assessing the penetration of bacteria.

$$\text{Penetration} = \frac{\text{number of bacteria passing through the filter}}{\text{challenge number of bacteria}}$$

Obviously for high-risk organisms which have low ID50 values it is very important that all appropriate precautions are taken.

9.14 Safety cabinets

Microbiological safety cabinets are designed to capture and retain airborne particles released in the course of certain manipulations and hence protect the laboratory worker from infections which may arise from inhaling them.

There are three kinds or classes of safety cabinet. Again as you might anticipate, there are some differences between the fine details of the specifications of different classes of safety cabinets. Nevertheless, the standards set are very similar in many countries. Each country imposes its own specifications. For example in the UK, these cabinets should conform to British Standard 5726. Below we give a general description of the various classes of cabinets that are available and indicate when they should be used. Note that these cabinets are variously described as Class 1, 2 and 3 cabinets or Class I, II or III cabinets.

9.14.1 Class I cabinet

A Class I cabinet is shown in Figure 9.3. Air is drawn in from the open front over the working area, it is then filtered to remove infectious particles.

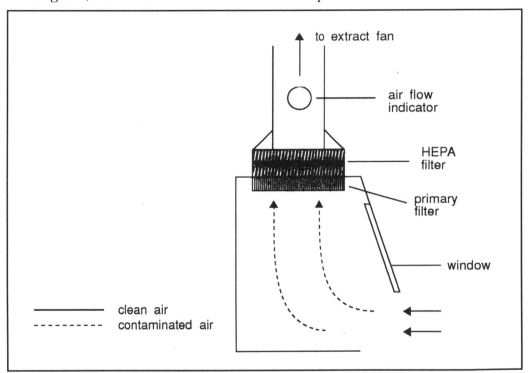

Figure 9.3 Class I microbiological safety cabinet, showing general design and airflow.

Typically, Class I cabinets are used with low-risk micro-organisms (EFB Class 2).

9.14.2 Class II cabinet

An example of a Class II cabinet is shown in Figure 9.4. Most of the air is recirculated through filters, some is dumped into the room and is replaced by air which is drawn through the open front. Class II cabinets protect the work within the cabinet from external contamination and provide some protection to operators.

In the UK, BS5726 has recently (from 1st January, 1993) been revised so that Class II cabinets dumping air into a laboratory must include dual in-line exhaust HEPA filters. This uprated standard, BS5726 (1992), is likely to be adopted across the EC.

This type of cabinet is used for medium-risk micro-organisms (EFB Class 3). Note that extensive precautions are taken to prevent the escape of organisms.

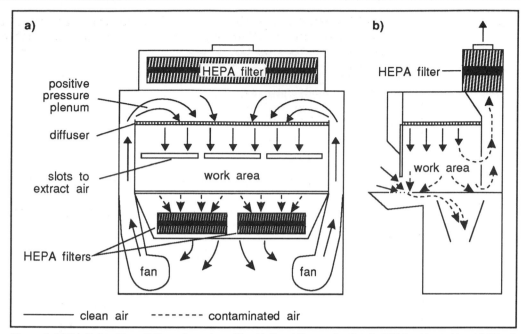

Figure 9.4 The Baker Class II Type 2 version of the National Cancer Institute's NCI-I cabinet. About 30% of the air is recirculated and 70% dumped after filtration. Air in the plenum has been filtered and is under positive pressure (courtesy of the Baker Company).

9.14.3 Class III cabinet

The cabinet is totally enclosed, the operator works via gloves attached to ports at the front. Air is filtered as it leaves the cabinet (see Figure 9.5). These cabinets are suitable for use in handling high-risk micro-organisms (EFB Class 4).

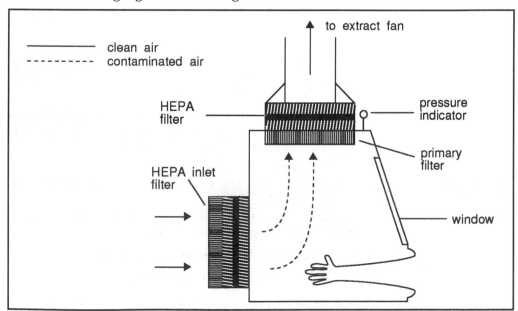

Figure 9.5 Class III microbiological safety cabinet, showing general design and airflow. Note that the micro-organisms and the worker are kept completely apart.

SAQ 9.7
Which class of safety cabinet should be used when manipulating the following bacteria or viruses? (You may need to look up the risk (hazard) group of the organisms in the Appendix).

1) *Salmonella typhi*.

2) *Corynebacterium diptheriae*.

3) Lassa fever virus.

4) non-pathogenic *Escherichia coli*.

9.15 High efficiency particulate air (HEPA) filters

HEPA filters are made from glass fibre paper which is approximately 60 μm thick. Fibres vary from 0.4-1.4 μm in diameter. The filter is constructed by folding the sheet into a pleat within a box unit, see Figure 9.6.

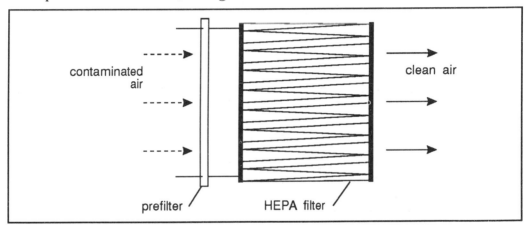

Figure 9.6 HEPA filter.

Biological safety cabinet HEPA filters are usually protected by coarser prefilters which remove dust and other particles down to about 5μm. These coarse filters are cheaper than HEPA filters and act to prolong their lives.

Even a good filter however, may let through 3 out of 100 000 organisms, so when hazardous organisms are being worked with and the exhaust air is vented into areas frequented by people two filters, in series, are used.

SAQ 9.8
1) Why are the figures obtained for air and surface contaminations in laboratories important?

2) Why might it be of value to employ two bacteriological filters in series when (EFB Class 3) organisms are cultured?

3) State whether the following are true or false:

   a) a Class I cabinet is fully enclosed;

   b) Class II cabinets must be used when working with (EFB Class 4) organisms.

9.16 Treatment of waste material

It is the responsibility of all laboratory workers to ensure that no infected material should ever leave the laboratory and become a risk to other workers or to the general public. Such material should be treated to render it safe. The practical methods of treating contaminated laboratory waste are:

- chemical disinfection;

- sterilisation by autoclaving;

- sterilisation by incineration.

The practical details of using these materials and procedures are provided in the BIOTOL text, '*In vitro* Cultivation of Micro-organisms'.

The choice of the method must be determined by the nature of the material to be treated. Figure 9.7 shows possible routes for the treatment of infected material.

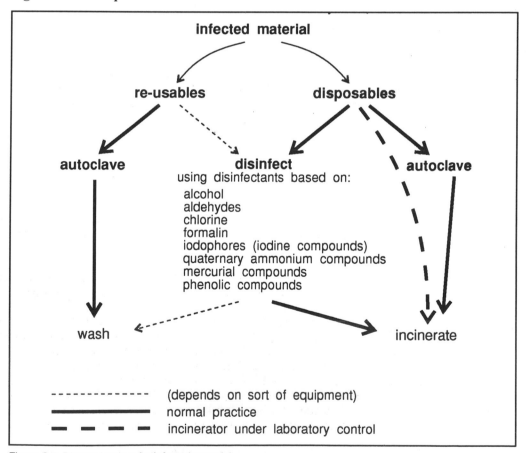

Figure 9.7 Disposal routes for infected material.

SAQ 9.9   Name four chemicals on which disinfectants may be based.

9.17 The special issues of animal cell cultures

So far we have predominantly discussed the safety issues concerned with the culture of micro-organisms. It is, however, important to consider animal cell cultures since they may present problems of safety. Currently animal cell lines are used to produce vaccines and many 'immortalised' (hybridoma) cell lines are used to produce monoclonal antibodies. Increasingly cell lines such as Chinese Hamster Ovary (CHO) cells are used to produce a variety of human and other mammalian proteins. These cell lines particularly lend themselves to genetic manipulation.

We will not review the origins of these cell lines in detail here (this aspect is covered in the BIOTOL text, 'In vitro Cultivation of Animal Cells'). We will, however, give a brief review of the major stages in producing cell lines as these have some bearing on the safety issues associated with such cell lines.

9.17.1 Methods used in the culture of animal cells

Initially animal cell cultures were of much importance to the study of how viruses attack human and animal cells. The technique of growing cells *in vitro* involves the use of animal cells of different organs. Generally, pieces of tissue are treated with trypsin solution to obtain single cells. Then a drop of this suspension is placed onto a flat surface (for example, in a Petri dish or on the inside of a flattened bottle of flask). The cells adhere to the surface and will grow if supplied with a nutrient solution containing about 10 amino acids and 10 growth factors or vitamins as well as salts, glucose and bicarbonate solution in equilibrium with 5% CO_2 in the gas phase. Often 5% calf or foetal calf serum is added.

primary culture

At this stage, the cells of this **primary culture** may still be contaminated with bacteria and viruses. Viruses can be detected by abnormalities in the growth of the cells and antibiotics such as penicillin and streptomycin are added to stop bacterial growth.

The cells will multiply until they occupy the whole surface (confluent cells). A few cells are then transferred to a sterile bottle with fresh sterile medium. Note that the cells only grow as a monolayer on the surface of the culture vessel.

The overall properties of the cells generally remain the same over many generations. Mutations may, however, occur and a particular mutant may become dominant.

established cell lines

These cells derived from normal cells cannot be subcultured indefinitely. After some 40-50 transfers the growth rate drops and the cells change their appearance: from the characteristic, normal euploid (diploid) pattern to an aneuploid stage, whereby numerous chromosomes and defective chromosome fragments appear. Then the cells die. Malignant tissue cells, however, give rise directly to cell lines that have an indefinite life span. If they have been grown for over 40 subcultures they may be considered as **established cell lines**. Malignant cells are often able to grow in suspension and much higher cell densities can be achieved than can be produced with normal cells.

Human viral vaccines may be produced by growing viruses on normal cells such as human embryo fibroblasts with finite life and non-tumorigenic characteristics. However, only limited quantities of vaccine can be produced (there are exceptions, for example, the large-scale production of Foot-and-Mouth Disease vaccine with the aid of baby hamster kidney cells).

Since malignant cells are able to grow in suspension at quite high densities, and are 'immortalised' (able to be sub-cultured indefinitely), increasing use is made of these cell lines for vaccine production. In this way unlimited quantities of antibodies may be produced.

Antigen-specific hybridomas or monoclonal antibodies are obtained by fusing myeloma (tumour) cells with lymphocytes of mice, for instance, which have been repeatedly immunised with an antigen of interest. Hybridomas may be grown in suspension for the production of monoclonals for commercial diagnostic uses.

In summary, animal cell cultures play an important role in the production of:

- human vaccines: Polio, Measles, Mumps, Rubella, etc;
- veterinary vaccines: Food-and-Mouth Disease, Rabies;
- monoclonal antibodies: for immune diagnosis, pregnancy tests, diagnosis of tumours, HIV tests, etc;
- human proteins: erythropoietin.

9.17.2 Safety considerations in using animal cells

oncogenes

Health complications have not been reported in laboratory work during 30 years of experience with cell lines. Nevertheless, animal cells contain proto-oncogene sequences in the cellular DNA. These can, in principle, be activated spontaneously under certain conditions and become carcinogenic or tumorigenic (eg by tumorigenic substances or retrovirus infections). Transformed cells may cause transplantable tumours in sensitive test animals but very seldom in humans in the case of accidental injury. Also remember that animal cells may harbour latent viruses.

latent viruses

The recommendations, therefore, concerning the safe handling of cell cultures (tissue explants, primary cell cultures, continuous cell lines) of human and animal origin as well as cell products derived from them, are also concerned with activated oncogenes and the possibility that such cells/cell products may harbour latent viruses.

contamination from other sourses

In addition remember explants and primary cell cultures may be contaminated with micro-organisms originating from human or animal donors, eg viruses and mollicutes (formerly mycoplasma). Furthermore, cell cultures may be contaminated accidentally during manipulations or via contaminated media components such as sera. It is, therefore, useful to list recommendations for the safe handling of animal cell culture.

9.17.3 Risks and safety precautions inherent with work with animal cells and cell lines

Primary cell cultures

Since the origin of microbiological contaminations in primary cell cultures are not known, all manipulations involving primary cell cultures should take place under Containment Category 1 (UK Hazard Group 2, Netherlands PG2) conditions, **unless** there are indications that medium-risk (EFB3) or high-risk organisms (EFB4) are present as contaminants. In these cases, the work should be performed under Containment Category 2 (CC2) or Containment Category 3 (CC3) conditions respectively. This may be the case with cells derived from primates with possible retrovirus infections; CC3 is then indicated. Retroviruses may be amphotrophic, that is, they may grow in related species.

germ-free (gnotobiotic) animals

The presence of pathogenic contaminants can be excluded if the cells were derived from healthy donors of non-suspect origin, eg germ-free (gnotobiotic) laboratory animals. Handling does not require physical containment in such cases.

SPF animals

Cells of specific pathogen-free (SPF) laboratory animals may be used. Physical containment is then only dependent on the type of micro-organism which may be still be present.

Established cell lines

All established cell lines should be considered as having potential low-risk (EFB2) and should consequently be handled under CC1 conditions, unless the cell lines are suspected of being contaminated with medium- or high-risk organisms. In these cases, the work should be carried out under CC2 or CC3 conditions. However, if the presence of pathogens in established tumour cell lines can be excluded, the cell line can be handled under GILSP conditions.

9.17.4 Risks involved with cell products

The risks connected with work with cell products from animal cell cultures (eg supernatants of cells, ascites and amnion fluids) may be classified as described below.

Cell products containing cells.

They should be handled according to the physical containment prescribed for the cells or cell lines from which they are derived.

Cell-free products

When cells are absent and therefore only possible microbiological contaminations play a role, the following physical containment levels should be considered:

- cell-free products which do not carry any microbiological contaminants;

  no physical containment is required, eg for mouse monoclonal antibodies in the absence of human pathogens;

- cell-free products which are contaminated with a known pathogenic micro-organism;

  these cell-free cellular products should be handled under the same physical containment as the contaminating micro-organism. When the pathogen has been eliminated from the cell products or inactivated, no physical containment for further handling is required. It is necessary to have the elimination or inactivation validated by standardised methods for every batch, for instance:

chemical inactivation:	standard determination of the residual concentration of the inactivating agent employed;
physical inactivation of retroviridae:	determination of the absence of reverse transcriptase activity.

We have summarised the risk class and safety precautions to be taken with animal cells, established cell lines and cell products in Table 9.9.

Animal cells/cell products	Risk class and safety precautions	
	Laboratory	Pilot plant
primary cell cultures		
unknown contaminants	low-risk (EFB2)	Containment Category 1
contains known pathogens	according to classification of pathogen	
pathogens are absent	harmless (EFB1)	GILSP
established cell lines/tumour cell lines		
unknown contaminants	low-risk (EFB2)	Containment Category 1
contains known pathogens	according to classification of pathogen	
pathogens are absent	harmless (EFB1)	GILSP
cell products		
contains viable cells	according to risk class of cell culture	
cell-free:		
original culture free from pathogens	harmless (EFB1)	GILSP
original culture with known pathogens	according to classification of pathogen	
validated and controlled inactivation of known pathogen	harmless (EFB1)	GILSP

Table 9.9 Summary of the risk classes of primary and established animal cell lines. EFB1 = European Federation of Biotechnology, Class 1 = harmless, GILSP = good industrial large-scale practice.

SAQ 9.10

1) Why are continuous (established) cell lines, although derived from oncogenic cell cultures, not considered dangerous?

2) What are gnotobiotic animals?

3) What are Mollicutes?

Summary and Objectives

In this chapter we have examined the biosafety issues involved with the use of natural micro-organisms in biotechnology. We have discussed the means by which the body may be harmed by pathogenic bacteria and the classification of micro-organisms into categories according to their hazard. As we saw, a pathogen must possess a number of properties which are not found in harmless species (saprophytes and commensals) in order to pose a threat to us. These facts should be kept in mind when the possibilities of creating pathogens by inserting pathogenic genes into harmless hosts are discussed in Chapter 10. We also discussed the safety issues involved in using animal cell lines.

Now that you have completed this chapter you should be able to:

- appreciate that micro-organisms are all around us in the environment;
- define the terms saprophyte, commensal, pathogen and virion;
- understand the significance of virulence and ID50;
- describe the defence mechanisms of the skin, the gastrointestinal tract and the respiratory tract;
- explain the properties that are necessary for an organism to be pathogenic;
- understand the difference between food poisoning and food infection;
- describe EFB Class 1,2,3 and 4 risk groups and equate these with UK ACDP hazard groups and the PG groups of The Netherlands;
- differentiate between containment levels used with different risk groups;
- name two instruments which can be used to monitor air contamination;
- describe two methods for monitoring surface contamination;
- describe how to test the efficiency of an air filter and define the term penetration in this context;
- sketch the three classes of safety cabinet;
- describe a HEPA air filter;
- sketch a flow chart to describe the treatment of contaminated waste;
- explain the risk classes of processes involving the use of animal cells.

Safety and the genetic manipulation of organisms

Safety and the genetic manipulation of organisms

10.1 Introduction

The undoubted advantages that may be accrued from genetically manipulating organisms are, to some extent, counteracted by the potential hazards which may arise from carrying out genetic manipulations. We may divide these hazards into two groups:

- those which threaten those who handle genetically modified organisms or carry out genetic manipulating procedures;

- those which offer a threat to the public at large.

The former of these two groups can be considered as the hazards associated with the contained uses of genetically manipulated organisms. The latter group are mainly those hazards associated with the release of manipulated organisms.

This text is primarily concerned with the laboratory procedures for genetically manipulating organisms, especially micro-organisms.

EC directives

From the onset you must realise that the contained use of genetically manipulated systems is the subject of both international and national regulations. Currently there are no universally accepted rules governing the genetic manipulation and use of micro-organisms although there are many similarities between the various regulations. For example the EC has produced two directives (Official Journal of the European Communities, L117, Vol 33, 8th May, 1990). One of these is concerned with the contained (laboratory) use of genetically manipulated organisms, the other specifies procedures for the deliberate release of genetically modified micro-organisms.

These directives are being used to modify national regulations and guidelines by the various EC member states.

The discussion of the regulation of the contained uses of genetically modified micro-organisms provided here is based upon the current systems operating within the UK and in The Netherlands. You will notice some significant differences in the procedures used but the experience this provides will give you a good understanding of the issues involved. We have divided the chapter into two main sections. In the first part, we discuss the assessment of hazards associated with the genetic manipulation of micro-organisms. This is done more-or-less from the standpoint of answering the questions - 'If I am intending to genetically manipulate a micro-organism, how do I evaluate the hazards this may present and what level of containment do I need to use?'

In the remaining parts of the chapter, we will discuss the EC directive on the contained use of genetically modified organisms with particular emphasis on micro-organisms. The deliberate release of genetically modified systems is largely beyond the scope of this text. This latter aspect of the regulation of the use of genetically manipulated organisms is more appropriately dealt with in the context of the use of such systems in the various business sectors and is examined in the relevant BIOTOL texts covering the application of biotechnology. Our strategy here is to provide you with experience of hazard assessment, relevant to laboratory study, not to provide you with complete details of the regulations.

The BIOTOL text, 'Biotechnology Source Book: Safety, Good Practice and Regulatory Affairs', gives a more extensive treatment of the safety issues associated with genetic manipulation.

10.2 Introduction to risk (hazard) assessment

In Chapter 9 we considered the hazards of naturally-occurring micro-organisms.

Π What happens to these considerations if we take a micro-organism from a harmless group (Hazard Group 1) and introduce genes that are foreign to that organism and that modify it so that it is no longer harmless?

Obviously we must alter the hazard group classification as appropriate and handle the organism accordingly.

licencing arrangements

Commission Genetic Modification

ACGM

Before we deal with the specifics of risk assessments associated with recombinant DNA technology, we need to provide you with some information concerning the regulatory framework in which such assessments are made. Each country has its own set of such regulations and authorities which monitor the application of these regulations. For example, in The Netherlands all recombinant DNA projects are regulated by the Nuisance Act. Facilities in which such work is carried out require a licence. The licence is given by the Community Council and is based on the advice of the Commission Genetic Modification (CGM). The Commission classifies each project into a risk group and advises on the safety measures to be implemented. A parallel arrangement operates in the UK. In the latter case, the Advisory Committee on Genetic Manipulation (ACGM) fulfils a similar function to the Commission Genetic Modification in The Netherlands.

Biosafety and Biological Safety Officers

The safety conditions of the workers in laboratories and industries involved in biotechnology are regulated by the ARBO law (Arbeidsomstandighederwet) in The Netherlands whilst in the UK this falls within the Health and Safety at Work Act. In both cases, the responsibility of the director of the institute or industry regarding the protection of workers is specified. In both cases, a Safety Officer with special training in recombinant DNA safety must be appointed. In The Netherlands the position carries the title Biosafety Officer (BSO) whilst in the UK the usual title is Biological Safety Officer.

In both countries, each project involving recombinant DNA has to be risk assessed and placed into a risk (hazard) category. This assessment is usually first carried out in-house by those wishing to carry out the project and the outcome of this assessment reported to the appropriate body (eg CGM in The Netherlands, ACGM in the UK). These bodies judge each proposal on merit and may advise granting approval of the project. Other countries, especially EC states and the USA, have analogous arrangements.

We believe it is important for you to understand the principles involved in carrying out a risk assessment on a proposal to use recombinant DNA technology. Unfortunately there is no universally accepted system in operation. It would become confusing to you to try to explain all of the systems that are in operation. We will use one system to show you the kind of strategy involved in evaluating a project. This is the system currently operating in the UK. You should be assured, however, that the adopted strategy is paralleled by the schemes operating in other states.

10.2.1 ACGM guidelines on risk assessment for proposals involving genetic manipulation

In the UK the Advisory Committee on Genetic Manipulation (ACGM) has produced a series of guidelines regarding genetic manipulation issues. We will consider their ACGM Note 7, 'Guidelines for the categorisation of genetic manipulation experiments'. ACGM Note 7 applies to laboratory cloning mainly in prokaryotes and lower eukaryotic organisms and to the genetic manipulation of plant cells. It is primarily concerned with the adverse consequences to those involved in genetic manipulation and also to those not directly involved.

The ACGM categorisation allows the consideration of possible risks associated with experiments under three different headings:

- access;

- expression;

- damage.

The scheme is a system of ranking, which allows the assignment of numbers to these risks and gives an overall score for each proposed experiment, which may be related to the level of containment required. We should keep in mind that the scheme is limited and the final figure can be no more accurate than the weakest component in the calculation. Values for access, expression and damage for certain host/vector combinations are given in Tables 10.1, 10.2, 10.3 and 10.4 and represent relative probabilities per unit bacterium. We will examine each of these tables in detail later. A difference of 10^3 is used between levels as the assessment of risks is applied to a wide range of experiments. In individual instances, it would be expected that more precise figures should be derived.

10.2.2 Access

access factor   The access factor is a measure of the probability that a manipulated organism, or the DNA within it, will be able to enter (gain access to) the human body and survive there. As you might imagine a value of 1 represents a situation where the majority of bacteria are expected to be able to enter the human body and survive there, 10^3 represents the chance of this occurring in 1 in 10^3 bacteria. Other access factors are also possible (see Table 10.1). As we saw in Chapter 9 we need to consider the different routes of infection involved depending on the organism used.

⨅   As an example, what access factor would you expect wild-type *Escherichia coli* and *Haemophilus influenzae* to have?

Wild-type *E. coli* and *H. influenzae* are examples of organisms which we expect to be able to invade and survive within humans and as such have access factors of 1. Table 10.1 shows the access factors for a number of different host/vector combinations.

In addition access factors for different host/vector combinations can be specified. Host/vector systems with an access factor of 10^{-6} are known as disabled, whilst those warranting an access factor of 10^{-9} as especially disabled.

Host/vector combination	Access factor
known ability to colonise humans eg	
wild-type *Salmonella* wild-type *Escherichia coli* *Haemophilus influenzae* *Staphylococcus aureus*	1
little known ability to colonise humans eg	
Streptomyces species *Bacillus subtilis* or non-colonising variants of a colonising organism eg *E. coli* K12; *E. coli* B & C	10^{-3}
disabled host/vector systems whether laboratory constructed or naturally occurring	$10^{-6} - 10^{-9}$
genetically manipulated DNA in tissue culture cells introduced as DNA which does not have the ability to infect or otherwise transfer to other cells.	10^{-12}

Table 10.1 Access factors of various host/vector combinations.

As you might expect, the properties of the vector are relevant to the consideration of access. The principle criterion for plasmid-based systems is that the plasmid should not be self-transmissible and should not be, or should be inefficiently mobilised by other plasmids. The plasmid pBR322 lacks the genes which encode proteins involved in mobilisation and is inefficiently mobilised by, for example, the F plasmid. However as it has the *nic* site at which proteins act, it can be mobilised if it coexists with a plasmid like *Col* E1 which can provide these proteins. The plasmid pAT153 in contrast lacks both the mobilisation proteins (*mob*⁻) and the *nic* site and so is absolutely defective for mobilisation. A list of 'key' vectors and *E. coli* hosts is shown in Table 10.2. Where a derivative of any of the listed vectors has been formed from other listed vectors it may also be regarded as disabled when used in an appropriate host, assuming that the derivative does not possess an enhanced ability to be mobilised, or to mobilise. This list is for guidance and is therefore not exclusive and local genetic manipulation safety committees may assign other vector/host systems into these categories as appropriate.

disabled vectors

Obviously mobilisation of a plasmid which encodes harmful products into a bacterial strain which survives well in human beings could allow the escape of damage-causing plasmids into the environment and is an important consideration. The use of suitable non-mobilisable plasmid vectors will therefore offer a significant safety margin.

This area is very complex so in this chapter we shall limit our discussions to plasmid vector/*E. coli* host systems, although similar categorisations are available for other host vector systems such as, for example, bacteriophage λ-based vectors, M13 vectors, cosmid vectors, yeast and *Bacillus subtilis* vectors.

Table 10.2 gives examples of acceptable relative values for your information, do not feel you have to learn them!

Vector plasmids	Host	Access factor
pBR322, pSC101	recombination deficient strains of *E. coli* K12	10^{-6}
pAT153, pACYC184, pUC series, pMTL series	*E. coli* K12	10^{-6}
pAT153, pACYC184 pUC series, pMTL series	especially disabled strains of *E. coli* K12 ie MRC 1, 7, 8, 9 X1776	10^{-9}
mob⁻ derivatives of plasmid incompatibility groups F, P, Q, W and X	*E. coli* K12	10^{-6}
mob⁻ derivatives of plasmid incompatibility groups F, P, Q, W and X	especially disabled strains of *E. coli* K12	10^{-9}
pBR313, pMB9, pAC134, pWT111, pWT121, pWT131, pOP213-13, pOP95-15	recombination deficient strains of *E. coli* K12	10^{-6}
pBR327, pBR328, pWT211, pWT221, pWT231	especially disabled strains of *E. coli* K12	10^{-9}

Table 10.2 Access factors associated with disabled *E. coli* host-vector systems.

You should realise that in this fast moving area, many new plasmids are being created. Thus it is important that you seek advice from the appropriate authority in your area; they publish recommendations for the classification of recombinant techniques involving vectors of the type described here.

SAQ 10.1

1) What is meant by the term access factor?

2) If an access factor of 1 is given, what does this mean?

3) Provide an example from Table 10.1 of a bacterium with an access factor of 1 and of another with an access factor of 10^{-3}. (Try to do this without looking back at Table 10.1)

4) Is a host/vector system with an access factor of 10^{-9} :

    a) disabled?

    b) especially disabled?

5) What properties of a plasmid vector are considered particularly relevant under the consideration of access?

6) What is the access factor for each of the following host/vector combinations?

    a) pAT153 in *E. coli* K12;

        10^{-3}, 10^{-6} or 10^{-9}.

    b) pAT153 in *E. coli* MRC 1;

        10^{-3}, 10^{-6} or 10^{-9}.

10.2.3 Expression

expression
factor

The expression factor is a measure of the known or anticipated level of a protein product to be made from the inserted DNA. If the expression is 1, then the system is designed to produce its product at a maximum rate. For initial experiments, the exact level of expression will be unknown and may be assessed using the information in Table 10.3.

Experiment	Expression factor
deliberate in frame insertion of expressible DNA downstream of a promoter with the intention of maximising expression	1
insertion of expressible DNA downstream of a promoter with no attempt to maximise expression, eg random cloning of cDNA in a phage λ gtII vector	10^{-3}
insertion of expressible DNA into a site not specifically situated to facilitate expression	10^{-6}
non-expressible sequence, eg non-coding sequence or gene broken by introns	10^{-9}

Table 10.3 Anticipated values for expression of inserted DNA in an initial cloning experiment.

Once the properties of clones have been defined, the expression factor can be calculated more precisely.

Whether or not the protein is expressed as a fusion product combining two different proteins should also be considered. This will generally effect the damage analysis rather than the expression factor (see Section 10.2.4).

As a rough guide, the maximum level of expression achievable within a single bacterium will be about 10^7 molecules. To give you an idea what this means, consider the following example. For an intracellular 40kDa protein product if we assume that the expression level is 10^6 molecules per bacterium, this would be equivalent to approximately 20% of the soluble cell protein and would represent 16 mg of the molecule in l g of bacterial cell paste.

SAQ 10.2

Assume that we have the same level of expression (10^6 molecules per cell) as in the example given in the text but this time the protein that we are considering is 100kDa in size.

Read the following questions and ring the correct answers in the lists provided.

1) What fraction of the soluble protein in the bacterium do you expect to contain this protein?

   5%   10%   20%   40%   50%   60%

2) If one litre of bacterial culture yields 10g of cell paste when we centrifuge the liquid to collect the bacterial cells together, what quantity of the 100kDa protein would we expect to find?

   4mg   8mg   16mg   32mg   40mg   80mg   400mg

10.2.4 Damage

damage factor

The damage factor is a means of assessing the risk of a gene product causing ill health to a worker exposed to it. The assessment is linked to likely damage caused by the protein product and the levels required to elicit this response.

For example, certain proteins are toxic at very low concentrations and therefore would have a damage factor of 1. Others have an effect at higher concentrations and might have a damage factor which is much lower. Table 10.4 gives some examples of damage factors. A DNA sequence which for example was known to be non-coding satellite DNA could have a damage factor of 10^{-12}.

Example	Damage factor
expression of a toxic substance or pathogenic determinant under conditions where it is likely to have a significant biological effect, eg ricin	1
expression of a biologically active substance which might have an effect if it were delivered to a target tissue	10^{-3}
expression of a biologically active molecule which is very unlikely to have a deleterious effect or, for example, where it could not approach the normal body level (as a rough guide up to 10% of the normal body level), eg human insulin	10^{-6}
use of a gene sequence where any biological effect is considered highly unlikely either because of the known properties of the protein or because of the high levels encountered in nature, eg human globin	10^{-9}
no foreseeable biological effect, eg non-coding DNA sequence	10^{-12}

Table 10.4 Some recommended damage factor values.

∏ Can you list some other factors which might effect the activity of the protein product?

Other considerations include:

* whether or not the protein which is produced is in a biologically active state;
* does it have requirements for glycosylation;
* is it effected by secondary modification or renaturation?
* is it going to be present in a site within the body where it can exert its effect? For example if it is produced in an organism which colonises the gut;
* will it be active where it is produced or will it be able to reach the target organ if this is different?

SAQ 10.3

Let us assume the worst possible instance, in which all of the *E. coli* in the gut are replaced by recombinant *E. coli* which are expressing a foreign polypeptide at the maximum theoretical rate. Furthermore, let us assume that these molecules are stable in the gut and may be delivered to their nearby target site. If there are 2×10^9 viable *E. coli* in the human gut and these each produce 10^7 foreign protein molecules each how many foreign protein molecules would we expect to be produced at any one time?

Whether this foreign protein had an effect on the body or not would depend on its mode of action. Some proteins for example are already found in the body at high concentrations, others might be foreign and act as powerful toxins.

10.2.5 Relationship between the risk assessment factors and containment level

When a particular project is being assessed the overall risk assessment value is calculated by multiplication of the individual figures obtained for access, expression and damage. The overall score can then be related, with the help of Table 10.5 to the suggested containment level

Overall risk assessment score	Containment level*
10^{-15} or lower	Containment level 1
$10^{-12} - 10^{-14}$	Containment level 2
$10^{-9} - 10^{-11}$	Containment level 3
$10^{-6} - 10^{-8}$	case by case
10^{-5} or greater	Containment level 4

Table 10.5 Relationship between the risk assessment value and the containment level. *Note that these are UK Containment Levels. In the EFB categorisation, Containment Level 1 equates with no containment requirement other than basic microbiological techniques; Containment Level 2 = EFB Containment Category 1 = FIN 1 in The Netherlands; Containment Level 3 = EFB Containment Category 2 = FIN 2 and so on (see Chapter 9).

SAQ 10.4

Consider the following examples and allocate to each the appropriate levels for access, expression, damage and containment in the grid below.

1) Cloning of an expressible human globin gene downstream of and in frame with an efficient promoter, in a plasmid vector which is to be used with wild-type *E. coli*.

2) Cloning of the same expressible human globin gene into a pUC series plasmid in order to optimise expression, using host *E. coli* MRC1.

3) Cloning of an expressible human insulin gene into a *mob⁻* derivative of a plasmid in incompatibility group Q in an attempt to maximise expression and in *E. coli* K12.

4) Cloning of non-coding mouse satellite DNA into pBR322 in recombination deficient *E. coli* K12.

Example	1)	2)	3)	4)
access factor				
expression factor				
damage factor				
containment level				
EFB containment category				

10.3 Genetically modified organisms - the European Community viewpoint

It is all very well one country having regulations regarding the use of genetically modified organisms but if these modified organisms are released into the environment in one country as a result of their continued use, they may reproduce and spread, crossing national frontiers and thereby affect other countries. In order to bring about the safe development of biotechnology throughout the European Community (EC), there has been an attempt to establish common measures for the evaluation and reduction of the potential risks arising in the course of operations involving the contained use of genetically modified micro-organisms.

10.3.1 The EC directive on the contained use of genetically modified micro-organisms

EC Council
Directive
The EC Council Directive of 23rd April, 1990 on the contained use of genetically modified organisms is published in the Official Journal of the European Communities, L117, Volume 33, 8th May, 1990. The purpose of the directive is to 'lay down common measures for the contained use of genetically modified micro-organisms with a view to protecting human health and the environment'. Progressively, national regulations are being adjusted to fit with these supra-national directives. In common with many such documents the legal wording of the EC Directives can be somewhat off-putting! The following sections may therefore be a little difficult but are worth persevering with. In this and other directives, the definitions which are used and the exemptions to the directive are of great importance to its scope and meaning.

Our first task is therefore to consider these.

10.3.2 Definitions.

A **micro-organism** is defined as 'any microbial entity, cellular or non-cellular, capable of replication or transferring genetic material'.

A **genetically modified micro-organism**, which we shall abbreviate to GMMO, is defined as 'a micro-organism in which the genetic material has been altered in a way that does not occur naturally by mating and/or natural recombination'.

In order to indicate techniques of genetic modification the following (non-exhaustive) list is given:

- recombinant DNA techniques using vector systems;

- techniques involving the direct introduction into a micro-organism of heritable material prepared outside the micro-organism including micro-injection, macro-injection and micro-encapsulation;

- cell fusion or hybridisation techniques to form new combinations of heritable genetic material which do not occur naturally.

Contained use means 'any operation in which micro-organisms are genetically modified or in which such organisms are cultured, stored, used, transported, destroyed or disposed of and for which physical barriers together with chemical and/or biological barriers, are used to limit their contact with the general population and the environment.

SAQ 10.5	Define the term micro-organism as used for *E. coli* using the EC Directive terminology.

SAQ 10.6	What does GMMO stand for?

SAQ 10.7	Give at least two examples of a technique said to involve genetic modification.

As we have seen, definitions are of great importance. Exemptions are also of great importance in regulatory affairs. They can on occasion make the difference between an experiment or procedure being subject to a regulation or not.

10.3.3 Exemptions

The following techniques are said not to result in genetic modification on condition that they do not involve the use of recombinant DNA or genetically modified organisms:

- *in vitro* fertilisation;
- conjugation, transduction, transformation or any other natural process;
- polyploidy induction.

The techniques of genetic manipulation which are excluded from the Directive so long as they do not involve the use of genetically modified micro-organisms as recipient or parental organisms are the following:

- mutagenesis;
- construction and use of somatic animal hybridoma cells;
- cell fusion of cells from plants which can be produced by traditional breeding methods;
- self-cloning of non-pathogenic naturally occurring micro-organisms which fulfil the criteria of Group 1 (see below) for recipient organisms.

Furthermore this directive does not apply to the transport of GMMOs or the storage, transport, destruction or disposal of GMMOs which have been placed on the market under community legislation, which includes an appropriate specific risk assessment.

At the time of writing, the applicability of some of these issues to the UK and The Netherlands is still under discussion.

10.3.4 Classification of GMMOs within the directive

For the purposes of the Directive, GMMOs must be classified into two groups, Group I and Group II.

Group I

Group 1 organisms have a long history of safe use and are considered to be safe when used under specific conditions. These criteria are similar in many ways to those we discussed in Chapter 9.

∏ Can you list five of the criteria that you think might be important?

The type of item you listed should compare with those indicated in Annex II of the Directive (shown in Table 10.6).

Group II

Those GMMO which do not satisfy the criteria of Annex II, ie are not in Group I, are said to fall into Group II.

Broadly speaking EFB Class 1, the Dutch PG Group 1 and the UK Hazard Group 1, which we discussed in Chapter 9 can be compared with Group I and the other EFB, PG and Hazard Groups as Group II.

Having classified a GMMO into the category Group I or II, the next step is to consider the type of operation or manipulation which is to be applied to the GMMO.

Criteria for classifying genetically modified micro-organisms in Group I

A **Recipient or parental organism**

— non-pathogenic;
— no adventitious agents;
— proven and extended history of safe use or built-in biological barriers, which, without interfering with optimal growth in the reactor or fermenter, confer limited survivability and replicability, without adverse consequences in the environment.

B **Vector/insert**

— well characterised and free from known harmful sequences;
— limited in size as much as possible to the genetic sequences required to perform the intended function;
— should not increase the stability of the construct in the environment (unless that is a requirement of intended function);
— should be poorly mobilisable;
— should not transfer any resistance markers to micro-organisms not known to acquire them naturally (if such acquisition could compromise use of drug to control disease agents).

C **Genetically modified micro-organisms**

— non-pathogenic;
— as safe in the reactor or fermenter as a recipient or parental organism, but with limited survivability and/or replicability without adverse consequences in the environment.

D **Other genetically modified micro-organisms that could be included in Group I if they meet the conditions in C above.**

— those constructed entirely from a single prokaryotic recipient (including its indigenous plasmids and viruses) or from a single eukaryotic recipient (including its chloroplasts, mitochondria, plasmids, but excluding viruses);
— those that consist entirely of genetic sequences from different species that exchange these sequences by known physiological processes.

Table 10.6 Annex II (from EC Council Directive of 23rd April, 1990 on the use of genetically modified micro-organisms).

Type A and Type B operations

Two types of operation are considered in the 1990 EC Directive on the use of genetically manipulated organisms. Type A and Type B operations.

Type A operations are 'any operation used for teaching, research, development, or non-industrial or non-commercial purposes and which is of a small scale (eg 10 litres or less)'.

Type B operations are any operations other than Type A operations.

It should be noted that for Type A operations some of the criteria shown in Annex II may not be applicable in determining the classification of all GMMOs. In which instance a competent authority should ensure that the relevant criteria are met.

SAQ 10.8

What type of operation would the examples below be classified as?

1) An undergraduate class of students streaking colonies of a GMMO onto an agar plate.

2) Fermentation of 100 litres of *E. coli* producing human interferon.

10.3.5 The system of the directive

The Directive states that all appropriate measures should be taken to avoid all adverse effects on human health and the environment through the contained use of GMMOs. To this end, the user is obliged to carry out a prior assessment of the risks that may occur taking into account the parameters which are relevant from Annex III of the Directive. Annex III is shown in Table 10.7

You should see many parallels between the list included in Table 10.7 and the criteria used in the UK system of risk assessment described in Section 10.2.

A record of this assessment should be kept by the user and made available to the competent authority if appropriate (see Section 10.2).

As you can see some of the points are fairly obvious and would be included in the UK ACGM Note 7 assessments.

Safety assessment parameters to be taken into account, as far as they are relevant, in accordance with Article 6 (3)

A   Characteristics of the donor, recipient or (where appropriate) parental organism(s).

B   Characteristics of the modified micro-organism

C   Health considerations

D   Environmental considerations

A   **Characteristics of the donor, recipient or (where appropriate) parental organism(s)**

— names and designation;
— degree of relatedness;
— sources of the organism(s);
— information on reproductive cycles (sexual/asexual) of the parental organism(s) or, where applicable, of the recipient micro-organism;
— history of prior genetic manipulations;
— stability of parental or of recipient organism in terms of relevant genetic traits;
— nature of pathogenicity and virulence, infectivity, toxicity and vectors of disease transmission;
— nature of indigenous vectors:
  sequence,
  frequency of mobilisation,
  specificity,
  presence of genes which confer resistance;
— host range;
— other potentially significant physiological traits;
— stability of these traits;
— natural habitat and geographical distribution. Climatic characteristics of original habitats;
— significant involvement in environmental processes (such as nitrogen fixation or pH regulation);
— interaction with, and effects on, other organisms in the environment (including likely competitive or symbiotic properties);
— ability to form survival structures (such as spores or sclerotia).

B   **Characteristics of the modified micro-organism**

— the description of the modification including the method for introducing the vector-insert into the recipient organism or the method used for achieving the genetic modification involved;
— the function of the genetic manipulation and/or of the new nucleic acid;
— nature and source of the vector;
— structure and amount of any vector and/or donor nucleic acid remaining in the final construction of the modified micro-organism;
— stability of the micro-organism in terms of genetic traits;
— frequency of mobilisation of inserted vector and/or genetic transfer capability;
— rate and level of expression of the new genetic material. Method and sensitivity of measurement;
— activity of the expressed protein.

Continued/...

Table 10.7 Annex III (from EC Council Directive of 23rd April, 1990 on the use of genetically modified micro-organisms).

C   **Health considerations**

—   toxic or allergenic effects of non-viable organisms and/or their metabolic products;
—   product hazards;
—   comparison of the modified micro-organism to the donor, recipient or (where appropriate) parental organism regarding pathogenicity;
—   capacity for colonisation;
—   if the micro-organism is pathogenic to humans who are immunocompetent:
        a) diseases caused and mechanisms of pathogenicity including invasiveness and virulence;
        b) communicability;
        c) infective dose;
        d) host range, possibility of alteration;
        e) possibility of survival outside of human host;
        f) presence of vectors or means of dissemination;
        g) biological stability;
        h) antibiotic-resistance patterns;
        i) allergenicity;
        j) availability of appropriate therapies.

D   **Environmental considerations**

—   factors affecting survival, multiplication and disseminations of the modified micro-organism in the environment;
—   available techniques for detection, identification and monitoring of the modified micro-organism;
—   available techniques for detecting transfer of the new genetic material to other organisms;
—   known and predicted habitats of the modified micro-organism;
—   description of ecosystems to which the micro-organism could be accidentally disseminated;
—   anticipated mechanism and result of interaction between the modified micro-organism and the organisms or micro-organisms which might be exposed in case of release into the the environment;
—   known or predicted effects on plants and animals such as pathogenicity, infectivity, toxicity, virulence, vector of pathogen, allergenicity, colonisation;
—   known or predicted involvement in biogeochemical processes;
—   availability of methods for decontamination of the area in case of release to the environment.

Table 10.7 (Continued) Annex III (from EC Council Directive of 23rd April, 1990 on the use of genetically modified micro-organisms).

SAQ 10.9

What are the four principle risk assessment headings to be considered prior to the use of a GMMO if the EC view point is adopted? (Do this without referring to Table 10.7).

10.3.6 Containment requirements for Group I and II GMMOs

Containment requirements for Group I GMMOs

The containment requirements for Group I organisms are set out in Article 7 of the Directive which is shown in Table 10.8.

∏   Compare these requirements with the requirements described in Chapter 9, for good microbiological practice and for the handling of 'safe' micro-organisms.

As you can see they have a similar basis.

For genetically modified micro-organisms in Group I, principles of good microbiological practice, and the following principles of good occupational safety and hygiene, shall apply:

i)   to keep workplace and environmental exposure to any physical, chemical or biological agent to the lowest practicable level;

ii)   to exercise engineering control measures at source and to supplement these with appropriate personal protective clothing and equipment where necessary;

iii)   to test adequately and maintain control measures and equipment;

iv)   to test, when necessary, for the presence of viable process organisms outside the primary physical containment;

v)   to provide training of personnel;

vi)   to establish biological safety committees or subcommittees as required;

vii)   to formulate and implement local codes of practice for the safety of personnel.

Table 10.8   Article 7 of the EC Council Directive of 23rd April, 1990 on the use of genetically modified micro-organisms.

Containment requirements for Group II GMMOs

In addition to Article 7 other containment measures should be applied to Group II GMMOs; they are set out in Annex IV of the Directive, which is shown in Table 10.9.

10.3.7 Activities which require authorisation or notification

For the purposes of the directive, the two Groups of GMMOs and the two types of operations can be split into four categories.

- Type A operations with Group I organisms - IA operations;

- Types B operations with Group I organisms - IB operations;

- Type A operations with Group II organisms - IIA operations;

- Type B operations with Group II organisms - IIB operations.

first use requires notification

The first use of an installation for an operation involving GMMOs is considered to be an activity for which notification of the authorities is required prior to commencement of the work.

conditions apply for subsequent use

Thereafter, Articles 9, 10 and 11 and Annex V of the Directive specify the type of notification which is required for each of the four categories of operations. They also describe the time lengths which in the case of Group IB and IIA operations must be waited before, in the absence of any indication to the contrary from the authorities, work can commence, or, in the case of Group IIB operations, the maximum length of time which the authorities will take to communicate their decision on whether or not the work may proceed.

Containment measures for micro-organisms in Group II

The containment measures for micro-organisms from Group II shall be chosen by the user from the categories below as appropriate to the micro-organisms and the operation in question in order to ensure the protection of the public health of the general population and the environment.

Type B operations shall be considered in terms of their unit operations. The characteristics of each operation will dictate the physical containment to be used at that stage. This will allow selection and design of process, plant and operating procedures best fitted to assure adequate and safe containment. Two important factors to be considered when selecting the equipment needed to implement the containment are the risk of, and the effects consequent on, equipment failure. Engineering practice may require increasingly stringent standards to reduce the risk of failure as the consequence of that failure becomes less tolerable.

Specific containment measures for Type A operations shall be established taking into account the containment categories below and bearing in mind the specific circumstances of such operations.

Specifications	Containment (EFB) categories		
	1	2	3
1) Viable micro-organisms should be contained in a system which physically separates the process from the environment (closed system):	yes	yes	yes
2) Exhaust gases from the closed system should be treated so as to:	minimise release	prevent release	prevent release
3) Sample collection, addition of materials to a closed system and transfer of viable micro-organisms to another closed system, should be performed so as to:	minimise release	prevent release	prevent release
4) Bulk culture fluids should not be removed from the closed system unless the viable micro-organisms have been:	inactivated by validated means	inactivated by validated chemical or physical means	inactivated by validated chemical or physical means
5) Seals should be designed so as to:	minimise release	prevent release	prevent release
6) Closed systems should be located within a controlled area:	optional	optional	yes, and purpose-built
a) Biohazard signs should be posted	optional	yes	yes
b) Access should be restricted to nominated personnel only	optional	yes	yes, via airlock
c) Personnel should wear protective clothing	yes, work clothing	yes	a complete change
d) Decontamination and washing facilities should be provided for personnel	yes	yes	yes
e) Personnel should shower before leaving	no	optional	yes
f) Effluent from sinks and showers should be collected and inactivated before release	no	optional	yes
			Continued/...

Table 10.9 Annex IV (from EC Council Directive of 23rd April, 1990 on the use of genetically modified micro-organisms).

Specifications	Containment (EFB) categories		
	1	2	3
g) The controlled area should be adequately ventilated to minimise air contamination	optional	optional	yes
h) The controlled area should be maintained at an air pressure negative to atmosphere	no	optional	yes
i) Input air and extract air to the controlled area should be HEPA filtered	no	optional	yes
j) The controlled area should be designed to contain spillage of the entire contents of the closed system	optional	yes	yes
k) The controlled area should be sealable to permit fumigation	no	optional	yes
7) Effluent treatment before final discharge:	inactivated by validated means	inactivated by validated chemical or physical means	inactivated by validated chemical means

Table 10.9 (Continued) Annex IV (from EC Council Directive of 23rd April, 1990 on the use of genetically modified micro-organisms).

10.3.8 Additional provisions of the Directive

Other Articles follow which discuss public consultation procedures, confidentiality, information on safety planning and possible emergency measures to be taken in the event of an accident but these are beyond the scope of this chapter.

10.4 Concluding comments

In this chapter we have examined two different ways of regulating the contained use of genetically manipulated organisms. In the UK, access, potential for expression and potential damage for each proposed genetic manipulation are assessed before an overall risk assessment factor is decided upon. The outcome of this overall risk assessment is to place the project into one of the hazard group categories. In contrast, the EC regulations, which are likely to be mirrored by subsequent UK procedures, places genetically modified micro-organisms into one of two groups on the basis of the history of safe use, the nature of the vector, genetic insert involved, survivability and pathogenicity. In the EC procedure, for each of these EC GMMO groups, the scale of operation and type of use permits further (Types A and B) classification to be made. We have learnt how the user of a GMMO is required to carry out a risk assessment based upon the donor and recipient organisms characteristics, any special characteristics of the modified organism and on health and environmental considerations. The regulations operating within the states which fall within the jurisdiction of the EC will also be finely-tuned to become consistent with the EC directives.

In all systems, there is a requirement for records of the assessment to be kept. Likewise, the outcome of the risk assessment places the project into a hazard category which in turn governs the nature of the containment which must be put in place. While systems differ in their methodology, they both seek to minimise risks from the genetic manipulation and uses of such manipulated systems.

Finally we remind you that here we have not examined the other important EC Directive on the deliberate release of genetically manipulated organisms. This directive (published in the Official Journal of the European Communities, L117 Vol B3, 8 May, 1990) mainly becomes effective when the manipulated organism is put to practical use (eg in food manufacture or, in agriculture or in environmental management). It is best dealt with within the context of these applications. The reader is, therefore, referred to the relevant BIOTOL texts dealing with biotechnological innovations in a variety of business sectors. This Directive is also discussed in the BIOTOL text, 'Biotechnology Source Book: Safety, Good Practise and Regulatory Affairs'.

Summary and objectives

In this chapter, we have examined two approaches (UK and EC) to carrying out risk assessments on the contained use of genetically manipulated systems. From time-to-time we have referred to other micro-organisms. These systems differ in methodology but share common objectives in providing for safety for workers using genetically modified organisms and the community at large.

Now that you have completed this chapter you should be able to:

- explain the importance of risk assesment in genetic manipulation;

- define and understand the terms access, expression and damage;

- describe the properties of a plasmid vector which are particularly relevant under the consideration of access;

- anticipate expression factor values for cloned DNA;

- describe some of the considerations relating to the biological activity of a protein;

- allocate access, expression and damage factors to proposed simple experiments;

- relate given risk assessment factors to an appropriate containment level;

- understand the basic principles of the EC Council Directive of 23rd April, 1990 on the contained use of genetically modified micro-organisms;

- use this Directive to define micro-organisms, genetically modified micro-organism and contained use;

- demonstrate awareness of the importance of the exemptions to this regulation;

- describe the criteria used to classify micro-organisms into Group I or Group II;

- define Type A and Type B operations;

- show awareness of the safety parameters which should be taken into account during risk assessment of genetically modified micro-organisms;

- understand the importance of different containment categories.

Responses to SAQs

Responses to Chapter 2 SAQs

2.1   The correct answer is 1). If you were incorrect you should read through the following definitions.

An auxotroph is a mutant micro-organism that cannot grow on minimal medium but requires the addition of some compound such as an amino acid (eg Leu, His) or a vitamin. Thus a Leu-requiring strain for example is referred to as *leu* ⁻. A prototroph is a micro-organism which can grow on minimal medium, Streptomycin, is a commonly used antibiotic with antibacterial properties. A sensitive organism is killed by streptomycin and is referred to as SmS. The results reported in Table 2.1 indicate that organism A requires histidine and leucine, can use galactase and is sensitive to streptomycin.

2.2   The answer is 3) for the following reasons.

1) This is incorrect, sinces the initial bacteria are both y^+ and z^+ they would be able to make their own y and z and hence would all grow on such a medium without DNA uptake.

2) This is incorrect, although DNA uptake into the bacteria will be selected for (only y^+ z^+ transformed bacteria will be able to grow) you would not be able to tell if the frequency of uptake of y was similar to that of z or not. By using media lacking both additives, we are only able to detect doubly transformed organisms.

3) Yes this is the correct answer. The frequencies of y^+, z^+ and y^+z^+ transformants can be assessed and compared. If the frequencies of these were all similar then it can be assumed that genes encoding y and z synthesis are close together. It is however important to use only very small amounts of DNA to ensure that only one transformation event occurs per cell!

2.3   When the cell [AB] enlarges, the repressor concentration will halve. Plasmids are selected at random for replication. Thus we might envisage that a cell may become [AAB] or [ABB]. As the cell continues to grow then the fall in repressor concentration allows for further replication. Again the selection of the plasmid which will be replicated is random. Obviously in cells with two copies of A and one of B, there is a greater chance that A will be replicated. Thus such a cell would become [AAAB] and, when the cell divides one daughter cell would become [AA] the other [AB]. This process would be repeated at each cell cycle. [AA] cells would only produce [AA] products. [AB] cells would produce [AA] and [AB] cells. In other words the proportion of [AB] cells would decline and the culture would become predominantly [AA]. If, on the other hand plasmid [B] had been the first to be replicated, then the culture would have become predominantly [BB].

We can represent this in the following way:

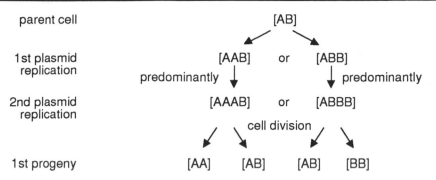

parent cell [AB]

1st plasmid replication [AAB] or [ABB]
predominantly ↓ ↓ predominantly

2nd plasmid replication [AAAB] or [ABBB]
cell division

1st progeny [AA] [AB] [AB] [BB]

Thus depending on the chance selection of plasmid for replication during the first cell cycle, either plasmid [A] or plasmid [B] will become dominant and the other plasmid will be, eventually, lost.

2.4

1) True, the ability to grow in the absence of arginine (Arg) means that the colonies can make their own arginine (hence prototrophic, arg^+).

2) False, the recipient bacteria are rec^- so any introduced DNA will not be able to recombine and insert into the recipient cell chromosome.

3) True, F' factors formed by faulty excision contain the arg^+ allele from the donor. This F' factor is transferred during conjugation to yield arg^+ recipients.

2.5

1) and 4).

Explanation:

1) a) Mixing phage with excess bacteria decreases the likelihood of more than one phage infecting each bacterial cell.

b) Removing the Ca^{2+} prevents any new phages from infecting transduced cells.

c) Selective medium prevents non-transduced bacteria from growing.

4) The only bacteria to grow will be those containing trp^+ transducing phage. Mixing phage with a host that is already lysogenic for a mutant phage Pl means that active Pl will not be able to lyse these cells. This is because the lysogenic P1 phage will produce a repressor which will switch off P1 lytic genes (see Section 2.7.2). Transducing phage which contain only bacterial DNA will be able to introduce their DNA and transduce the bacterium.

The following reasons can be given for 2) and 3) as unsatisfactory approaches. In 2), since we are infecting a few cells with many P1 phages, there is a high probability that cells which receive trp^+ particles will also receive lytic P1 phage particles and would, therefore, be lysed. In 3), although some cells would only receive trp^+ particles others would receive lytic P1 particles. These would multiply inside their hosts, be released and, in the presence of Ca^{2+}, infect other cells (including those containing trp^+). These cells too would subsequently be lysed. Thus both 2) and 3) would not be satisfactory for isolating transduced bacterial cells.

2.6

① ③ ④ ⑤ ⑥ ⑦

Responses to Chapter 3 SAQs

3.1

1	a	b	c	d	e	ⓕ	g
2	a	b	ⓒ	d	e	f	g
3	a	b	ⓒ	d	e	f	g
4	a	ⓑ	c	d	e	f	g
5	a	b	c	d	e	f	ⓖ
6	a	b	c	ⓓ	e	f	g
7	a	b	c	ⓓ	e	f	g
8	a	b	c	d	ⓔ	f	g
9	ⓐ	ⓑ*	c	d	e	f	g

Note b* will precipitate RNA not DNA.

3.2   Absorbance = 0.2.

1 absorbance unit corresponds to 50 µg ml^{-1} of double-stranded DNA.

0.2 absorbance units correspond to 0.2 x 50 µg ml^{-1} = 10 µg ml^{-1} of double-stranded DNA.

The DNA was initially diluted by 100 fold therefore the concentration was 100 x 10 µg ml^{-1} = 1000 µg ml^{-1} of double-stranded DNA. Thus you should have ringed answer 1).

3.3   a) use method 2).

b) use method 4).

c) use method 5).

d) use method 1).

e) use method 3).

3.4   1) False, there are 3.

2) False, Class II are used.

3) False, this is the cut site for *Hin* dIII.

4) True.

5) False, methylation here will prevent restriction.

6) True.

7) True, some also have tetramers and pentamers.

8) True.

9) True.

10) False, *Sma* I and *Hae* III do, *Bam* HI does not.

If you have problems with these draw out the double-stranded sequences by referring to Tables 3.4 and 3.5. It is worth persevering to understand these, as they are fundamental to the processes of genetic engineering.

3.5 The answer is band e.

Described below are the details of all the bands on the gel:

Lane 1 - linear pieces of λ *Hin* dIII DNA.

Lane 2 - uncut plasmid DNA: b is the CCC DNA which moves fastest; a is the OC DNA.

(Note that both of these bands represent the entire 3000 bp plasmid and that due to its structure its migration is not proportional to the linear DNA markers in lane 1).

Lane 3 - band c is linear DNA 3000 bp long. (Note that there is only one *Sal* I site).

Lane 4 - band d is the *Bam* HI fragment which is found if you look at the plasmid map and move clockwise round from the *Bam* HI site at nucleotide 640 around to the next site at position 2740. This represents a total of 2100 bp. This DNA fragment as expected is found between the 2322 and 2027 bp λ *Hin* dIII markers.

Band f represents the piece of DNA between nucleotides 2740 and 2900 and is hence 160 bp in length and migrates less than the smaller 125 bp λ *Hin* dIII size marker.

Band e represents the piece of DNA between the nucleotides 2900 and 640. It is therefore 100 + 640 = 740 bp long and is the smallest piece of DNA which includes gene X.

3.6
1) Possible. As there is no direct positive reaction for plasmid containing an insert, maybe only 1 in 100 clones will contain an insert, the rest would probably contain reformed vectors without gene X inserted (re-ligated vectors).

2) Not possible. The first part of the statement '*Bam* HI and *Bcl* I termini are compatible' is true as is the fact that 'bacteria containing plasmids with inserts can be selected by their tetracycline resistance and sensitivity to ampicillin' (vectors without an insert will be both tetracycline and ampicillin resistant). **But,** the ligation of *Bam* HI and *Bcl* I termini together yields a hybrid:

which cannot be cut with either enzyme (note it would however be cut by *Sau* 3A or *Mbo* I which cleave ↓GATC).

3) We could use this method, as colonies carrying the vector with an insert are tetracycline sensitive and ampicillin resistant. Colonies containing the vector alone are resistant to both antibiotics.

4) We could use this method. Colonies containing vector and insert are tetracycline resistant and ampicillin sensitive. If we use the enzyme *Bcl* I we can cut out the insert as a means of confirmation.

All of the strategies described could yield *E. coli* carrying gene X, but some of these strategies have limitations. For example, in 1), we have no way of ensuring that *E. coli* acquiring antibiotic resistance have also acquired gene X. We have to go through the process of extracting the plasmid DNA from these antibiotic resistant cells and checking the DNA fragments produced after restriction enzyme digestion.

In 2) we can positively select for *E. coli* cells that contain plasmids which have disrupted *amp* R genes. We might assume that such a plasmid has had the *Bam*HI fragment inserted into it, but we could not be sure of this without further analysis. This is possible but not so straightforward as we would have to use an alternative restriction enzyme to carry out the DNA restriction enzyme fragmentation analysis (see 2) above).

3) looks quite straightforward but again we would really need to carry out a restriction enzyme fragment analysis to confirm that the disrupted *tet* R gene really was carrying the desired insert. 4) is also satisfactory.

You will see from this discussion that there is usually more than one approach to introducing new genes. Do not be too concerned if you did not get all of this question right; by the time you have completed this text you will have a lot more experience and will be able to make good judgements about such strategies.

Responses to Chapter 4 SAQs

4.1

1) *S. coelicolor* = $10^7/2 = 0.5 \times 10^7$ per μg added DNA.

   S. hygroscopicus = $25/2 = 12.5$ per μg added DNA.

2) *S. coelicolor* = 10^7 transformants from $10^9 \times 0.25$ protoplasts = 25 protoplasts needed per transformant (or 0.04 transformants per viable protoplast).

   S. hygroscopicus = 25 transformants from $10^9 \times 0.25$ protoplasts = 10^7 protoplasts needed per transformant (or 10^7 transformants per viable protoplast).

   This shows that only a small proportion of cells are potentially transformable. This sub-population of particularly susceptible cells is a common feature of many transformation systems. The strain which appears less easy to transform may require different optimum conditions or possess some other problem like a restriction system that degrades incoming DNA.

4.2

1) The cells will be in an actively growing log phase at this OD.

2) By adding 100 mmol l^{-1} $CaCl_2$ for 20 minutes.

3) Ice $\rightarrow$ 42°C for 2 minutes $\rightarrow$ ice.

4) Any media that will select for the plasmid DNA markers, usually antibiotic resistance, are used for selection of transformants.

5) Hopefully, we will see colonies of *E. coli* resistant to the antibiotic growing on the selective Petri dishes. It would be normal to include controls (which are Petri dishes containing nutrient media with no antibiotic) to check cell viability. We might also plate out cells which had no exogenous DNA to score for spontaneous revertants resistant to the antibiotic.

4.3 10 000 plaques = 10 µl of suspension.
10 000 x 100 = number of plaques formed from entire volume of 1 ml.
So 10^6 plaque forming units per 0.5 µg of DNA which = 2×10^6 pfu per µg of vector DNA.

4.4 Possibly, the infection conditions for *E. coli* were not optimal. Thus, although 1×10^9 pfu may have been formed in the lysate, because unsuitable conditions were used to infect *E. coli* with these reconstructed phages, only a small proportion of the phages infected cells. Thus relatively few phages were produced. Alternatively, your ligation reactions were not 100% efficient and so much less than 0.5 µg of substrate DNA was actually available for packaging.

It is unlikely that you will be offered your money back! However you could try asking the manufacturers for a sample of their λ DNA just to check out their claims.

4.5 4; 3; 2; 1; 6; 5.
It would be normal to clean the DNA, perhaps by using phenol extraction followed by ethanol precipitation between steps 4-3 and 2-1 to allow resuspension in the optimal restriction and ligation buffers. Congratulations you have made a new vector.

4.6 The gene must be about 0.1 x 1/100 x 200 000 kb in length = 200 kb.

Only the YAC can hope to clone this gene intact.

Almost all YAC transformation protocols use protoplast (yeast) transformation. Although attempts are being made to use simpler protocols like the lithium acetate technique, the large size of these vectors makes it difficult to visualise how they may be introduced into cells without first removing the cell walls.

4.7 b) and c) and e). The *cos* site betrays its phage ancestors but its size is far too small to contain most of the essential phage functions. This suggests it is a cosmid. The selectable markers for *A. nidulans*, (*pyr*4 and *ans*l) show it can be used to transform fungi and ampR is the bacterial selectable marker. So it must also be a shuttle vector. We can see no yeast telomeric regions and so pCAP2 cannot be a YAC. The small, circular nature of the molecule and its ability to code for antibiotic resistance and origin of replication suggest it is a plasmid.

4.8 1) The entire construct must be 38 to 52 kb in size to be packaged (see Figure SAQ 3.8), therefore the maximum insert = 52-8.65 = 43.35 kb and the minimum insert = 38-8.65 = 29.35 kb.

2) We will lose the ability to select for transformants of *A. nidulans*. Even though the DNA may enter the cell/protoplast we lose the selectable marker. Of course, the deleted derivative could still be used as an *E. coli* cosmid but it would no longer be a shuttle vector. The cloning capacity (including minimum insert size) would increase, although this would really be of little practical significance.

4.9 1) A plasmid would be the obvious choice based on the small size of the insert. A phage with a small cloning capacity might also be used.

2) A YAC would be the only vector with a large enough capacity to take up such a huge gene intact.

3) Potentially any of these vectors could do this job. However, in practice this insert size would be far too small for most cosmids and for effective use of the YAC. We are left with a choice of a plasmid, although the insert is rather large for ease of handling. We could use a phage with a large cloning capacity like λ EMBL4.

Responses to Chapter 5 SAQs

5.1

1) The gene of interest may be on a *Bam* HI fragment that is too large for the capacity of the cloning vector.

2) We applied a total digest and therefore many of the fragments may be very small, perhaps even so short that they cannot be cloned in some vectors.

3) The gene of interest may have internal *Bam* HI sites so that it is cut and cannot be cloned as a single fragment. This may of course always happen but becomes more likely when using a total digest.

5.2

If we insert the respective numbers into our equation; 99% = p of 0.99:

$$N = \frac{\ln (1 - 0.99)}{\ln (1 - 45/1.4 \times 10^{5)}}$$

$= 1.4 \times 10^4$ clones.

Remember, we should have to screen many more clones than this if we have a vector with a smaller capacity. This is a very important consideration when we think of the work we have to put into screening a particular gene bank.

5.3

The advantages are:

1) we need to isolate fewer clones;

2) we preferentially isolate expressed (transcribed) genes;

3) we can often enrich for inducible sequences;

4) we can remove promoter and intron sequences.

The disadvantages are:

1) RNA isolation is less easy than DNA isolation and RNA is less stable than DNA;

2) there are more steps that can go wrong!

3) if we are interested in the promoter and intron sequences, a cDNA bank is not suitable;

4) the spatial arrangements of genes that are visible on large inserts in genomic banks are not visible in cDNA banks;

5) we can only isolate expressed genes.

Once more we return to the key point that we can only really choose our vector and bank when we have decided on a screening method for the gene of interest. Check again with Section 5.3 if these points are not clear.

5.4

Not necessarily. It is possible that reversion of the *arg⁻* gene in the host occurred without transformation of a wild-type gene. It is important to include a control which would allow detection and measurement of reversion. The simplest control would be to include in our original transformation experiment a batch of cells that were plated out on MM without any added DNA.

5.5

c, i, a, d, e, b, j, f, b, g, h.

5.6   There are a number of ways to check the identity of the amplified DNA:

1) look for cross-hybridisation with a suitable probe;

2) check the size of the amplified DNA with that expected from the separation of the primers;

3) restriction map the DNA;

4) sequence the DNA and compare with the known sequence.

5.7   1) complementation screening; 2) PCR; 3) immunological/protein screening; 4) nucleic acid hybridisation.

5.8   1) PCR would be the obvious first choice since the gene has been sequenced, primers can thus be made available and the technique is fast.

2) The complementation method is the best bet. Our lack of knowledge of the gene itself, the existence of auxotrophs and a host that we can transform would make this a feasible approach.

3) Nucleic acid hybridisation methods could use the fungal and mouse genes as probes - we can reprobe the same filters and avoid the extra steps involved in trying to make antibodies. PCR may be a possibility if strongly conserved sequences are apparent and we could try a variety of primers.

4) Each of the above methods would be possible although the complementation technique would involve complex cell culture methods. PCR could be used if strongly conserved DNA sequences existed. Nucleic acid probes could be designed if the degeneracy of the genetic code had conserved enough similarity at the DNA level between insulin genes from different organisms. The immunological/protein screening method would, however, probably be the method of choice.

Remember, there may be more than one screening strategy available for a particular job. Invariably, where sequence data are available, PCR is the easiest option. Where suitable probes can be made, the hybridisation methods are quickest. Complementation screening can be straightforward where transformation is efficient. In the end, the final justification of the chosen method is if it works!

Responses to Chapter 6 SAQs

6.1   1) G-C to A-T is a transition.

2) A-T to G-C is a transition.

3) G-C to T-A is a transversion.

4) GCTCGT to GCTCT is a deletion (G deleted).

5) GCTCGT to GCTCGGT is an insertion (G inserted).

6) GCTCGT to GCTTGTC is both a transition (C - T) and an insertion (C added).

7) This is an inversion; the sequence TCCGG has been inserted in the
                                                      AGGCC
opposite orientation.

Note that mutations 1)-6) described in the question can also be described as point mutations.

6.2

1) Conversion A-T to G-C is a transition and the mutation could be produced by nitrous acid.

2) Conversion of G-C to A-T is a transition and the mutation could be produced by either nitrous acid or hydroxylamine.

   Notice that hydroxylamine *only* converts G-C to A-T and not *vice versa*.

3) Mutation of G-C to T-A is a transversion, the mutation could be produced by hydrazine which is the only chemical named capable of producing a transversion.

4) Conversion of G-C to A-T is a transition, it can be produced upon treatment of single-stranded DNA with sodium bisulphite.

5) The mutation of ATCTGA to ATCGTGA is an insertion it could be produced by acridines, the only chemical mentioned that produces insertions.

6.3

Kunkel method	relies on destruction of DNA containing uracil
allows multiple rounds of mutagenesis without re-cloning	cyclic selection
allows strand selection *in vivo*	amber mutations
allows strand selection *in vitro*	thionucleotides
an advantage of the gapped-duplex method	only a small window of single-stranded DNA is available for the oligonucleotide to anneal to
dam⁻ strains	produce undermethylated template DNA
vectors capable of producing single-strand DNA	M13
an enzyme used in producing a mutant heteroduplex	DNA polymerase
an enzyme involved in excision of uracil from DNA	uracil glycosylase

If you had difficulty in answering this question, we suggest you read Section 6.4 again.

6.4

1) The lack of restriction enzyme sites and the need for random deletions impose choices such as gap-sealing mutagenesis, which is in this case the only suitable method. Producing nested deletions with DNase I usually relies on the use of restriction enzymes and so is unsuitable.

2) Random insertions can be produced by linker-insertion mutagenesis and linker-scanning mutagenesis. However, as the question does not state what size of deletions are needed, it is probably better to start with the much simpler technique of linker-insertion mutagenesis.

3) The method of choice here would be linker-scanning mutagenesis which locates important areas accurately and does not rely on the presence of restriction sites in the cloned DNA fragment.

4) Restriction sites at each end of the cloned DNA and the requirement for only approximate location makes the use of *Bal* 31 to create nested deletions attractive.

5) With only one restriction site available, digestion with *Bal* 31 and exonuclease III are excluded; we could use DNase I to produce nested deletions.

For 4) and 5) other methods such as gap-sealing mutagenesis could be used as well, but the use of restriction enzymes probably makes the use of the other techniques more efficient.

6.5 At first sight it may appear that the cysteine mutation is the most useful because of its increased reaction rate. However, you will see that alanine substitutions give rise to an enzyme that is far more resistant to oxidation. Thus alanine mutations retain far more activity in conditions where the oxidative stress is high. Thus the alanine mutant may prove to be more useful in the actual detergent. However, you will also notice that both the alanine and cysteine mutations produce proteases with low affinities for substrates. At the low substrate concentrations that proteases in detergents are required to work on this may also be an important consideration.

What you will learn from this question is that the problems encountered in protein engineering may not have a simple clear cut solution. In the case cited in the question, the mutations lead to changes in at least three parameters (reaction rate, affinity for the substrate and resistance to oxidation).

Amino acid substitutions may have many effects, all of which need to be characterised. These effects may also prove to be 'antagonistic' requiring a compromise solution to be reached. The data for this question are based on real data from mutagenesis of a protease (subtilisin) which is used in detergents.

Responses to Chapter 7 SAQs

7.1 To tackle this problem, you should begin by listing the information that you are certain of if you do not have the confidence to go straight into the question. Such information might be that we know the size of the whole piece of linear DNA must be 10 000 basepairs and that we know the digest information is complete as all the fragment sizes add up to 10 000 basepairs.

With the necessary information we can build up the restriction map as follows.

We can start with (for example) the 4200 basepair fragment. This is cut by *Hin* dIII to give two fragments of 400 basepairs and 3800 basepairs. This is seen below:

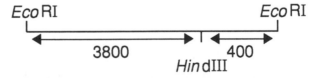

We know that the 3800 basepair fragment is also produced by digesting the 5000 basepair fragment with *Eco* RI. We also know that the 400 basepair fragment comes from digestion of the 2400 basepair fragment with enzyme *Eco* RI. We can add to our map as follows:

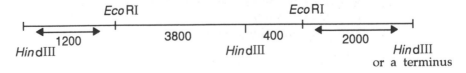

Carrying on this technique of looking for overlapping fragments we see that the 1200 basepair fragment is also produced by digestion with *Hin* dIII of the 2800 basepair fragment. This also allows us to locate the other end of the piece of DNA because the 1600 basepair fragment just added to our map is also produced by digestion of the 2600 basepair fragment with *Eco* RI. The map therefore looks like:

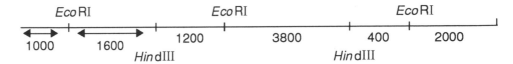

7.2   Remember that in this type of sequencing the presence of G and C residues is indicated by the occurrence of bands in both the G and A + G lanes or the C and C + T lanes respectively. Furthermore the gel is also read from the bottom of the gel upwards.

G	A + G	C + T	C	
				3′
		───	───	C
───	───			G
	───			A
	───			A
		───		T
		───		T
		───	───	C
───	───			G
───	───			G
───	───			G
		───		T
		───	───	C
───				A
		───	───	C
		───		T
───	───			G
				5′

7.3   Let us start by recapping on the equations for RNA titration analysis.

The fraction of cellular RNA homologous to the tracer is given by:

$H = S \times L1/L2$ where S is the slope of the titration graph. L1 is the length of the RNA being analysed and L2 is the length of the tracer.

We have the information necessary to solve this as follows:

$H = 0.0001 \times 2000/200$ which gives us 0.001.

We can now use the second equation to find the number of homologous RNA molecules per cell.

Number of molecules per cell = $(H \times R \times 6 \times 10^{23})/(F \times 339 \times L1)$

We know H and R and we can assume that all the cells express the RNA so we know F.

We have then:

Number of molecules per cell = $(0.001 \times 2 \times 10^9 \times 6 \times 10^{23})/(1 \times 339 \times 2000)$ which gives us 1 503 750 molecules per cell, which is a very high number (and quite unrealistic)! This is because the slope of the titration graph you were given was in fact quite high.

7.4 The correct order is: 3) isolate nuclei; 2) produce and label RNA *in vitro*; 4) isolate RNA from contaminating DNA and proteins; 1) hybridise to cloned DNA; 5) quantify the degree of hybridisation.

7.5 In this question you were given several tasks which you had to solve by the use of appropriate techniques. Let us look at this problem more closely.

1) You need to identify the protein product of a cloned cDNA. This is fairly simple as only one of the responses deals with proteins! You should know though that hybrid arrest of translation will give some indication of the protein product of a clone. So method c) is the correct choice. You may, however, also wish to get more information by expressing the clone in a heterologous system.

2) You need to know where in a plant a particular RNA sequence is expressed. This is an ideal opportunity for Northern blotting and hybridisation. RNA from different parts of the plant can be analysed by this technique. Thus method d) would be the correct answer. If the plant were small enough, or if only a particular part of the plant was of interest then *in situ* hybridisation could be used.

3) You need to know if the different levels of an mRNA found by Northern blotting are due to different transcription rates in different parts of a plant. The analysis of transcription *in vitro* will provide data on transcription rates. Nuclei isolated from different parts of the plant would have to be used. Thus the correct answer is d).

4) You need to subclone small, easily sequenced pieces of a large cDNA into a vector and need to know the location of restriction sites. Both sequencing and restriction mapping would provide this information but, as restriction mapping is easier and provides enough accuracy for this application, it is probably the best choice. Thus the correct answer is g).

5) You need to know the precise number of RNA molecules in a particular cell type. Only RNA titration analysis will provide this kind of precision data, so method b) is the correct choice.

6) As a precursor to 5), you need to know what percentage of cells in a mixed sample is expressing the RNA sequence. Of the techniques listed here only *in situ* hybridisation would allow analysis of individual cell types for expression of the RNA, which means that method e) would be the correct answer. If the cell types could be easily separated, then RNA titration analysis could also be used.

7) You need precise data on restriction enzyme sites in a piece of DNA. This type of precise information can only be gained by sequencing, restriction mapping is not accurate enough. Either dideoxy or chemical sequencing could be used. Thus both method f) and method h) could be used.

8)

8) You need to identify the position of methylated bases in a piece of DNA. As we saw earlier the chemical sequencing method allows the position of modified bases to be identified, whereas dideoxy sequencing only gives information in terms of the four usual bases, so here method h) must be chosen.

Responses to Chapter 8 SAQs

8.1

1) The 5' end of the DNA would be labelled.

2) 600 bp.

3) No band would be seen in lane 2 as there is no complementary RNA to protect the DNA from digestion.

4) 400 bp is the length of the hybrid formed.

The pattern of the gel bands would thus look like:

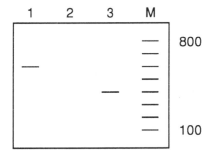

Go back over Section 8.2.2 if you have had difficulty with this SAQ.

8.2

1) Lane 1 just shows the position of the unreacted probe.

2) The band in lane 2 shows no extension because there is no RNA present to act as a template for the reverse transcription.

3) The primer extension product shows a length of 250 bases as this is the length of the reverse transcript, which includes the primer.

The pattern of the bands would look thus like:

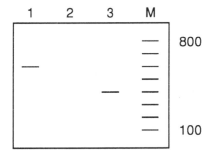

8.3

The correct order for the steps would be: 4; 5; 2; 1; 3.

You would begin by isolating full length clones that contain the sequences you are interested in (steps 4 and 5). You would then need to characterise the fragments produced from digesting the clone (step 2) before you could go on to look for expression patterns (step 1) and investigate what sequences were responsible for the pattern of expression (step 3).

We can now look at the questions.

1) What reporter genes are particularly useful in plants?

   β-glucuronidase is becoming a very widely used reporter gene in plants because it is a) present in very low levels in plants; b) easy to assay and c) can be used in histochemical reactions to visualise tissue specific gene expression for example.

2) What features might a piece of DNA have to ease manipulations?

   Amongst the desirable features would be small size so that the DNA fragment is not easily broken during manipulations and the presence of suitable restriction sites to allow subcloning into the vector containing the reporter gene at high efficiency. It might be useful if the restriction sites at each end of the DNA fragment were different so that the DNA could be cloned into the vector in a specific orientation.

3) What mutagenesis techniques might be used to identify promoter regions of interest?

   The answer to this question can really be found in Chapter 6, but the two topics are closely linked. However a technique such as deletion mutagenesis with nucleases is probably a good start as fairly large and controlled deletions can be made so that general areas can be identified.

4) What type of sequences may cause walks to split?

   The presence of repetitious DNA sequences may split the walk into several different directions.

5) What is the advantage of primer extension?

   Primer extension product allows the terminus to be mapped precisely as the product can easily be sequenced after being recovered from the gel.

8.4

1) b.

2) c.

3) a.

4) d.

Re-examine Section 8.5 if you found this SAQ difficult.

Responses to Chapter 9 SAQs

9.1

1) A pathogen is an organism which causes disease.

2) A saprophyte is an organism which metabolises dead organic matter.

3) A commensal is an organism which benefits by living within another organism without causing disease.

9.2

1) Fatty acids, salt and the low pH of the healthy skin have an antibacterial action.

2) The gastrointestinal tract is protected against invading pathogens by the hydrochloric acid in the stomach and furthermore the rapid transport of food through the tract and the enteric flora of the lower tract give little chance for the

development of most pathogens. The intestinal mucous layer containing secreted antibodies also plays an important role.

3) The respiratory surface has a mucous coating that traps most types of microbial life. The entrapped cells can then be killed by phagocytic cells. The mucous coating is swept up the respiratory passages by cilia, and it is coughed up and subsequently swallowed, thereby depositing potentially infectious material into the bactericidal contents of the stomach.

4) Phagocytic cells attach to intruding pathogens, pseudopodia extend from the phagocytic cell and fuse around the pathogen to form a phagocytic vesicle in which the invader is killed.

9.3

1) The influenza virus gets out of the body of an infected person by coughing and sneezing and is inhaled by new hosts.

2) A virion is a complete virus particle and consists of nucleic acid and protein.

3) Food poisoning is the illness caused by the ingestion of foods containing bacterial toxins and is generally not contagious.

4) Food infection is the contagious illness caused by foods in which a viable pathogen is present.

5) *Clostridium botulinum* - Yes;

   Staphylococcus aureus - Yes;

   The organism which causes diphtheria - Yes;

   The organism which causes tetanus - Yes.

See Sections 9.5-9.7 if you have had problems with these questions.

9.4

EFB Risk Classes 2,3,4 and ACDP Hazard Groups 2, 3 and 4 may be referred to as pathogens since they may cause human disease. This question emphasises the similarities between the EFB and ACDP systems.

9.5

1) Medium- and high-risk groups (EFB Class 3 and 4; UK Hazard Groups 3 and 4; Netherlands PG 3 and 4).

2) Medium- and high-risk groups (EFB Class 3 and 4; UK Hazard Groups 3 and 4; Netherlands PG 3 and 4).

3) All groups (EFB Class 1,2,3 and 4; UK Hazard Groups 1,2,3 and 4; Netherlands PG 1,2,3 and 4).

4) All groups.

5) High-risk groups (EFB Class 4; UK Hazard Group 4; Netherlands PG 4).

9.6

1) *Staphylococcus aureus* is in Hazard Group 2 (see Appendix 3).

2) Containment Level 2 should be used for work involving this organism. If you answer EFB CC1, or FIN 1, these are equally satisfactory.

3) A safety cabinet should be used if a culture of this organism has to be vigorously mixed. This prevents aerosols, resulting from the vigorous mixing, contaminating the workplace. *Staphylococcus aureus* may cause disease.

4) If the bacterial mass is to be disposed of it should be carefully placed in a robust container without spillage and sterilised, all other materials which have come into contact with the organism should also be sterilised.

5) Microbiological contamination is the presence of micro-organisms in places where they are not intended to be such as on work benches in a laboratory or on the worker's hands.

Infection is the invasion of the body by a pathogenic organism. For example if the contaminating micro-organisms on the worker's hands were pathogenic and entered a cut or abrasion, they could cause disease.

9.7

1) *Salmonella typhi* is a medium-risk organism (EFB Class 3, Hazard Group 3). Therefore, you should use a Class II cabinet.

2) *Corynebacterium diptheriae* is a low-risk organism (EFB Class 2). You might have been surprised to find it in this category. A Class I cabinet should be used.

3) Lassa fever virus is a high-risk organism. (EFB Class 4). It must be handled in a Class III cabinet.

4) Non-pathogenic *Escherichia coli* is regarded as harmless (EFB Class 1) and no containment cabinet is needed.

9.8

1) The figures obtained for air and surface contaminations in laboratories are important because they enable us to check that safe working practices are being followed.

2) It is of value to employ two bacteriological filters in series when EFB Class 3 (Hazard Group 3) organisms are cultured because even small numbers of organisms escaping through a single filter may be capable of causing a serious infection.

3) a) False. A Class I cabinet is not fully enclosed.

   b) False. A Class III cabinet must be used when working with EFB Class 4 organisms and not a Class II cabinet.

9.9

You could have chosen four disinfectants based on chemicals from the following list:

alcohol; aldehydes; chlorine; formalin; iodophores; quaternary ammonium compounds; mercurial compounds; phenolic compounds.

These types of compound can all be used to treat objects which are contaminated with micro-organisms (see Section 9.16, Figure 9.7).

9.10

1) Continuous cell lines have been used for 30 years without infecting the investigators. If established cell lines are free from pathogens, they may be considered as harmless. Moreover they are cultivated under such strict asepsis (avoiding entrance of unwanted micro-organisms) that good microbiological techniques are always fulfilled.

2) Gnotobiotic animals are germ-free laboratory animals (rats or mice).

3) Mollicutes were formerly named mycoplasma. The mollicutes family comprises mycoplasma, acholeplasma and ureaplasma. They are low-risk organisms. *M. pneumoniae* causes pneumonia in groups of people in close contact, eg soldiers in barracks, children in schools.

Responses to Chapter 10 SAQs

10.1

1) The access factor is a measure of the probability that a manipulated organism, or the DNA within it, will be able to enter the human body and survive there.

2) A value of 1 represents a situation where the majority of bacteria are expected to be able to enter a host and survive there.

3) Examples given in Table 10.1 of bacteria with an access factor of 1 are:

   wild-type *Salmonella*; wild-type *Escherichia coli*; *Haemophilus influenzae*; *Staphylococcus aureus*.

   Examples given in Table 10.1 of organisms with an access factor of 10^{-3} are:

   Streptomyces species; *Bacillus subtilis*; non-colonising variants of a colonising organism eg *E. coli* K12, *E. coli* B and C.

4) A host/vector system with an access factor of 10^{-9} is especially disabled (b).

5) The principle criterion for plasmid based systems is that the plasmid should not be self-transmissible and should not be, or should be inefficiently, mobilised by other plasmids.

6) The access figure for the following host/vector combinations are as follows:

   pAT153 in *E. coli* K12 $= 10^{-6}$;

   pAT153 in *E. coli* MRC1 $= 10^{-9}$.

10.2

1) The protein will comprise about 50% of the soluble protein in the bacterium. In our previous example a 40kDa intracellular protein is equivalent to 20% of the cells soluble protein. A 100kDa protein is 2.5 x this mass, therefore you would expect 2.5 x 20% of the soluble protein in the bacterium to be composed of this 100kDa protein.

2) 400mg (or 0.4g) of protein in 10g cell paste. The 40kDA protein represents 16 mg per 1g of cell paste. Therefore the 100kDa protein represents 2.5 x 16 mg per 1 g = 40 mg. However we are considering the paste obtained from a one litre culture, which we are told weighs 10 g so we must multiply our answer by 10.

10.3

We would expect $2 \times 10^9 \times 10^7 = 2 \times 10^{16}$ foreign protein molecules to be produced at any one time.

10.4

Example	1)	2)	3)	4)
access factor (A)	1	10^{-9}	10^{-6}	10^{-6}
expression factor (E)	1	1	1	10^{-9}
damage factor (D)	10^{-9}	10^{-9}	10^{-6}	10^{-12}
A x E x D	10^{-9}	10^{-18}	10^{-12}	10^{-27}
Containment Level	3	1	2	1
EFB Containment Category	2	BMT	1	BMT

Note: BMT = basic microbiological techniques.

10.5   A micro-organism is defined as 'any microbiological entity, cellular or non-cellular, capable of replication or transferring genetic material'.

10.6   GMMO stands for genetically modified micro-organism. GMMO, is defined as 'a micro-organism in which the genetic material has been altered in a way that does not occur naturally by mating and/or natural recombination'.

10.7   Techniques said to involve genetic modification include:

- recombinant DNA techniques using vector systems;

- techniques involving the direct introduction into a micro-organism of heritable material prepared outside the micro-organism including micro-injection, macro-injection and micro-encapsulation;

- cell fusion or hybridisation techniques to form new combinations of heritable genetic material which do not occur naturally.

10.8   1) An undergraduate class of students streaking colonies of a GMMO onto an agar plate would be performing a Type A operation.

2) Fermentation of 100 litres of *E. coli* producing human interferon would be a Type B operation.

10.9   The four principle risk assessment headings to be considered prior to the use of a GMMO if the EC view point is adopted are:

- characteristics of the donor, recipient or (where appropriate) parent organism(s);

- characteristics of the modified micro-organism;

- health considerations;

- environmental considerations.

Appendix 1 - Restriction endonucleases and their recognition sequences

This appendix includes a selection from the several hundred restriction enzymes that have been characterised. They are arranged in order of the sequences they recognise based on a system and nomenclature used by Boehringer Mannheim GmbH, Biochemica (PO Box 310-120, D-6800 Mannheim, Germany) - see also Kessler C, Höltke HJ (1986) Gene 47;1.

Recognition sequence	Restriction Enzyme	Recognition sequence	Restriction Enzyme
G↓Å°Å⁺TTĊ⁺	*Eco* RI	A↓AGCTT	*Eco* VIII
↓GÅ°TC	*Bce* 243	Å⁺↓AGĊ⁺TT	*Hin* dIII
↓GATC	*Bsp* AI	A↓AGCTT	*Hsu* I
↓GÅ°TC	*Cpf* I	GÅ°GĊ⁺T↓C	*Sac* I
GA↓TC	*Dpn* I	GA GĊ⁺T↓C	*Sst*
↓GÅ⁺TC	*Fnu* CI/*Hac* I	CA G↓Ċ⁺TC	*Pvu* II
↓GÅ⁺TĊ°	*Mbo* I	GG↓Ċ⁺C	*Bsu* RI
A↓GÅ°Ċ⁺T	*Bgl* II	GG↓CC	*Clt* I/*Fru* DI/*Sfa* I
A↓GATCT	*Nsp* MACI	GG↓Ċ⁺Ċ°	*Hae* III
G↓GATCC	*Ali* I	ᵀ/C↓GGC CᴬG	*Cfr* 1/*Cfr* 141
G↓GÅ°TĊ⁺Ċ°	*Bam* HI	ᵀ/C↓GGĊ⁺C ᴬG	*Eae* I
G↓GATĊ⁺C	*Bst* I	CG↓CG	*Fnu* DII
ᴬ/G↓GÅ°TC ᵀ/C	*Mfl* I	Ċ⁺G↓Ċ⁺G	*Tha* I
ᴬ/G↓GÅ°TĊ⁺ ᵀ/C	*Xho* II	G↓CGCGC	*Bss* HII
CGAT↓CG	*Nbl* I	CCGC↓GG	*Csc* I/*Gce* I/*Gce* GLI/*Sst* II
CGÅ°T↓Ċ⁺G	*Pvu* I	Ċ⁺CGC↓GG	*Sac* II
CGAT↓CG	*Rsh* I	cᴬ/C G↓C ᵀ/G G	*Nsp* BII
CGÅ°T↓Ċ°G	*Xor* II	TCG↓CGÅ°	*Nru* I
T↓GÅ°TĊ°A	*Bcl* I	AT GCA↓T	*Eco* T22/*Nsi* I
CATG↓	*Nla* III	CT GCA↓G	*Ali* AJI/*Bsp* I/*Cfl* I/*Sal* PI
GCATG↓C	*Pae* I	Ċ⁺T GCÅ⁺↓G	*Pst* I
GCATG↓Ċ°	*Sph* I	CT GCÅ⁺↓G	*Sfl* I
ᴬ/G CATG↓ ᵀ/C	*Nsp* 75241/*Nsp* HI	A↓Ċ⁺GT	*Mae* II
Ċ⁺↓CATGG	*Nco* I	GA CGT↓C	*Aat* II
AG↓Ċ⁺T	*Alu* I	TAC↓GTA	*Sna* BI
AG↓CT	*Mlt* I		

Recognition sequence	Restriction Enzyme	Recognition sequence	Restriction Enzyme
G$\overset{+}{C}$G↓C	*Cfo* I	G↓GTAC$\overset{+}{C}$	*Asp* 718
GCG↓C	*Fnu* DIII	C↓TAG	*Mae* I
G↓$\overset{+}{C}$GC	*Hin* P1 I	A↓CTAGT	*Spe* I
AGC↓GCT	*Eco* 47111	G↓CTAGC	*Nhe* I
GGCGC↓C	*Bbe* I	C↓CTAGG	*Avr* II
GG↓CGC$\overset{\circ}{C}$	*Nar* I	T↓$\overset{+}{C}$TAG$\overset{+}{A}$	*Xba* I
GG↓CGCC	*Nda* I	GTT↓A$\overset{+}{A}\overset{\circ}{C}$	*Hpa* I
G${A \atop G}$↓CG${T \atop C}$C	*Acy* I/*Ast* WI/ *Asu* III/*Hgi* DI	C↓TTAAG	*Afl* II
G${A \atop G}$↓$\overset{+}{C}$G${T \atop C}\overset{+}{}$C	*Aha* II	TTT↓AAA	*Aha* III/*Dra* I
TGC↓GCA	*Aos* I/*Fdi* II/*Mst* I	G↓G${A \atop T}$CC	*Afl* I
C↓$\overset{+}{C}$GG	*Hap* II	G↓G${A \atop T}$C$\overset{+}{C}$	*Ava* II
C↓CGG	*Mno* I	G↓G${A \atop T}$CC	*Bam* Nx/*Hgi* BI/ *Hgi* II/*Hgi* EI
$\overset{+}{C}$↓$\overset{\circ}{C}$GG	*Msp* I	GG↓${A \atop T}$CC	*Bam* 216
G$\overset{+}{C}$C↓GG$\overset{+}{C}$	*Nae* I	${A \atop G}$G↓G${A \atop T}$CC${T \atop C}$	*Ppu* MI
$\overset{+}{C}$↓$\overset{+}{C}\overset{\circ}{C}$GGG	*Cfr* 91	CG↓G${A \atop T}$CCG	*Rsr* II
CC$\overset{+}{C}$↓GGG	*Sma* I	C$\overset{\circ}{C}$↓${A \atop T}$GG	*Aor* I
T↓$\overset{\circ}{C}$G$\overset{+}{A}$	*Taq* I/*Tth* HBBI	$\overset{\circ}{C}\overset{\circ}{C}$↓${A \atop T}$GG	*Bst* NI
G↓TCGAC	*Hgi* CIII/*Hgi* DII/*Nop* I	↓$\overset{\circ}{C}\overset{+}{C}$${A \atop T}$GG	*Eco* RII
G↓T$\overset{+}{C}$G$\overset{+}{A}$C	*Sal* I	$\overset{\circ}{C}\overset{\circ}{C}$${G \atop C}$↓GG	*Bcn* I
GT↓${AT \atop CG}\overset{+}{A}$C	*Acc* I	CC↓${G \atop C}$GG	*Cau* II
C↓TCGAG	*Blu* I/*Ccr* I	G↓ANTC	*Fru* AI
C↓TCG$\overset{+}{A}$G	*Pae* R7	G↓$\overset{+}{A}$NT$\overset{+}{C}$	*Hin* fI
C↓T$\overset{+}{C}$G$\overset{+}{A}$G	*Xho* I	G↓GNCC	*Bsp* BII
C↓TCGAG	*Xpa* I	↓C$\overset{+}{C}$NGG	*Sso* 4711
TT↓CGAA	*Asu* II/*Fsp* II/*Mla* I	↓GTNAC	*Mae* III
ATT↓ATT	*Ssp* I	G↓GTNACC	*Asp* AI
G$\overset{+}{A}$T↓ATC	*Eco* RV	GC↓TNAGC	*Esp* I
CA↓TATG	*Nde* I	$\overset{m}{C}$ restricts at all 5-methylcytosines	*Ala* JA1I
GT↓AC	*Rsa* I		
AGT↓ACT	*Sca* I		

Symmetrical sequences are specified only for the 5′ ⟶ 3′ strand
Symbol N represents any nucleotide
$\overset{+}{A}$ and $\overset{+}{C}$ indicate that methylation of the bases inhibits the restriction enzymes
A and C indicate that restriction endonuclease activity is not affected by methylation of these bases

Appendix 2 - Selection of enzymes used in molecular biology and genetic engineering

This appendix is divided into DNA polymerases, RNA polymerases, nucleic and cleaving enzymes, nucleic acid joining enzymes, termini changing enzymes, changing DNA morphology and methylating enzymes.

Enzyme	Source	Activity	Main appliction
DNA polymerases			
DNA polymerase 1 (Kornberg polymerase)	E. coli	i) synthesises DNA 5' → 3', needs ssDNA and a primer with 3'-OH terminus; ii) 5' → 3', exonuclease from 5'-P terminus of dsDNA; iii) 5' → 3', exonuclease from 3'-OH terminus of ssDNA	labelling DNA by 'nick' translation synthesis of second strand of cDNA
DNA polymerase 1 (Klenow fragment)	E. coli	similar to i) and iii) of Kornberg polymerase above	second synthesis of cDNA and in site-directed mutagenesis; end-filling and labelling of dsDNA with 5' overhang; production of ssDNA probes by primer extension
T4 DNA polymerase	T4 infected E. coli	similar to Klenow fragment above	end-filling of dsDNA; gap-filling in site-directed mutagenesis
Taq DNA polymerase	Thermus aquaticus	similar to i) and iii) of Kornberg polymerase above	thermostability - finds use in DNA amplification and genome footprinting
micrococcal DNA polymerase λ DNA polymerase	Micrococcus luteus calf	alternatives to DNA polymerase 1	
AMV reverse transcriptase	avian Myeloblactosis virus pol gene in recombinant E. coli	i) RNA-dependent DNA polymerase 5' → 3', requires ssRNA (or DNA) and a primer with 3'-OH terminus; ii) exoribonuclease activity	first strand of cDNA synthesis; modification of chain termination; DNA sequencing
M-MuLV reverse transcriptase	Moloney munne leukemia virus pol gene in recombinant E. coli	an alternative to AMV reverse transcriptase	

RNA polymerases

Enzyme	Source	Activity	Main appliction
E. coli RNA polymerase	*E. coli*	DNA-dependent RNA polymerase	*in vitro* transcription, labelling of RNA
also other RNA polymerases with specificities for alternative promoters. Examples are: SP6 RNA polymerase (SP6 promoters); T3 RNA polymerase (T3 promoters); T7 RNA polymerase (T7 promoters).			

nucleic acid cleaving enzymes (other than restriction enzymes - see Appendix 1)

Enzyme	Source	Activity	Main appliction
ribonuclease A	Bovine pancrease	endonuclease, cleaves 3′ phosphodiester bond	removes RNA from DNA
ribonuclease H	*E. coli*	endoribonuclease, removes RNA from RNA-DNA hybrids	cDNA synthesis, removes poly A from mRNA
deoxyribonuclease 1	Bovine pancrease	endonuclease, cleaves 5′ to pyrimidines	removal of DNA from RNA preparation, introduction of 'nicks' into dsDNA
Bal 31 nuclease	Alteromonas espejiana Bal 31	sequential removal of nucleotides 3′ → 5′ of ssDNA or dsDNA	tailoring of dsDNA by removing unwanted nucleotides; restriction mapping
S1 nuclease	Aspergillus oryzae	endonuclease most active on ssDNA	removal of overhangs
mung bean nuclease	Vigna radiata	endonuclease	an alternative to S1 nuclease
T7 endonuclease	T7 infected *E. coli*	endonuclease most active on ssDNA	
exonuclease 1	*E. coli*	cleaves ssDNA 3′ → 5′	
λ exonuclease	*E. coli* λ lysogen	5′ → 3′ exonuclease active on dsDNA	removes 5′ overhangs
phosphodiesterase	various	hydrolyses DNA and RNA from 5′ end	

Enzyme	Source	Activity	Main appliction
nucleic acid joining enymes (ligases)			
T4 DNA ligase	*E. coli* λ lysogen	joins blunt-ended, coehesive-ended and nicked molecules, require 5'-P and 3'-OH termini	joining DNA
E. coli DNA ligase	*E. coli*		
T4 RNA ligase	T4-infected *E. coli*	joins ssDNA and ssRNA with 5'-P and 3'OH groups	joining ssRNA (DNA) molecules
changing the termini of nucleic acid molecules			
alkaline phosphatase	various	removes 5'-P groups from nucleic acids;	dephosphorylating dsDNA to prevent self-ligation
T4 polynucleotide kinase	T4-infected *E. coli*	adds P to 5'-OH of nucleic acids (uses ATP)	labelling of 5' termini prior to Maxim and Gilbert sequencing
poly(A) polymerase	*E. coli*	uses ATP	to label RNA with nucleotides
polynucleotide phosphorylase	various	adds ribonucleotides to 3'-OH termini of RNA molecules	makes artificial RNA molecules
mRNA guanyl transferase	various	produce 5'-cap structures on RNA	
changing the morphology of DNA			
topoisomerase I	calf thymus	relaxes supercoils of DNA by 'nicking'	
DNA gyrase topoisomerase II	bacterial	introduces negative coiling in CCCDNA	

DNA methylating enzymes*

Enzyme	Specificity	Enzyme	Specificity
M-*Afl* II	CTTAAGC(^{M6}A)	M-*Eco* PI	AG^{M6}ACC
M-*Alu* I	AG^{M5}CT	M-*Eco* RI dam	G^{M6}ATC
M-*Apa* I	GGG^{M5}CCC	M-*Ecol* RI	GA^{M6}ATTC
M-*Bal* I	TGG^{M5}CCA	M-*Fru* DII	^{M5}CGCG
M-*Bam* HI	GGAT^{M4}CC	M-*Hae* III	GG^{M5}CC
M-*Bvb* SIII	A^{M6}AG	M-*Hin* dII	^{M6}AAGCTT
M-*Bsp* RI	GC^{M5}CC	M-*Mbo* I	G^{M6}ATC
M-*Bst* VI	CTCG^{M6}AG	M-*Ngo* II	GG^{M5}CC
M-*Bsu* FI	^{M5}CCGG	M-*Sal* I	GTCG^{M6}AC
M-*Cfr* 9I	C^{M4}CCGGG	M-*Taq* I	TCG^{M6}A
M-*Dpn* II	C^{M6}ATC	M-*Xba* I	TCTAG^{M6}A
M-*Eco* dam	G^{M6}ATC		

* Positions of methyl groups on the bases are specified

Appendix 3 - Categorisation of micro-organisms according to hazard

Within the main body of the text, we have referred to the fact that micro-organisms may be categorised according to the extent that they represent a hazard to health. Four groups are identified - Hazard Group 1 being of lowest risk, Hazard Group 4 presenting the greatest hazard.

The extent and nature of the containment that needs to be implemented when using micro-organisms is governed to a large extent by the hazard group to which the micro-organism belongs.

Here we provide a list of organisms according to their hazard group. The list includes only organisms in Hazard Groups 2, 3 and 4. The names that have been used are those in common (but not universal) usage. Many have one or more synonyms.

The list provided here is based on the categorisation of pathogens by the UK, Advisory Committee on Pathogens ('Categorisation of Pathogens According to Hazard and Categories of Containment', HMSO, 1990).

In countries other than the UK, slightly different nomenclature may be used although the same (or very similar) criteria for assigning micro-organisms to 1 of 4 categories is often employed (see Chapter 9).

The list provided here should not be regarded as universally applicable for all time and for all places. As new knowledge emerges or as new control measures are developed, it becomes appropriate to re-categorise some organism. Furthermore, it is possible to develop attenuated or genetically modify strains of pathogens organism such that they may be re-categorised.

The reader should also be aware that specific conditions may apply to particular pathogens. For example, deliberate cultivation of the *Variola* (smallpox) virus is banned totally in some countries. A key aspect of working with micro-organisms is therefore, not only to work safely, but also be aware of the regulatory obligations that must be fulfilled.

BACTERIA, CHLAMYDIAS, RICKETTSIAS AND MYCOPLASMAS

Hazard Group 3

Bacillus anthracis

Brucella spp.

Chlamydia psittaci (avian strains only)

Coxiella burnetti

Francisella tularensis (Type A)

Mycobacterium africanum

Mycobacterium avium

Mycobacterium bovis (excl BCG strain)

Mycobacterium intracellulare

Mycobacterium kansasii

Mycobacterium leprae

Mycobacterium malmoense

Mycobacterium paratuberculosis

Mycobacterium scrofulaceum

Mycobacterium simiae

Mycobacterium szulgai

Mycobacterium tuberculosis

Mycobacterium xenopi

Pseudomonas mallei

Pseudomonas pseudomallei

Rickettsia-like organisms

Rickettsia spp.

Salmonella paratyphi A, B, C

Salmonella typhi

Shigella dysenteriae (Type 1)

Yersinia pseudotuberculosis subsp pestis (Y pestis)

Hazard Group 2

Acinetobacter calcoaceticus

Acintoebacter lwoffi

Actinobacillus spp.

Actinomadura spp.

Actinomyces bovis

Actinomyces israelii

Aeromonas hydrophila

Alcaligenes spp.

Arizona spp.

Bacillus cereus

Bacteroides spp.

Bacterionemia matruchottii

Bartonella bacilliformis

Bordetella parapertussis

Bordetella pertusis

Borrelia spp.

Campylobacter spp.

Cardiobacterium hominis

Chlamydia spp. (other than aivan strains)

Clostridium botulinum

Clostridium tetani

Legionella spp.

Leptospira spp.

Listeria monocytogenes

Moraxella spp.

Morganella morganii

Mycobacterium bovis (BCG strain)

Mycobacterium chelonei

Mycobacterium fortuitum

Mycobacterium marinum

Mycobacterium microti

Mycobacterium ulcerans

Mycoplasma pneumoniae

Neisseria spp. (spp .known to be pathogenic for Man)

Nocardia asteroides

Nocardia brasiliensis

Pasteurella spp.

Peptostreptococcus spp.

Plesiomonas shigelloides

Proteus spp.

Providencia spp.

Pseudomonas spp. (other spp known to be pathogenic for Man)

Hazard Group 2 (Cont)

Clostridium spp. (other spp. known to be pathogenic for Man)

Corynebacterium diphtheriae

Corynebacterium spp. (other spp. known to be pathogenic for Man)

Edwardsiella tarda

Eikenella corrodens

Enterobacter spp.

Erysipelothrix rhusiopathiae

Escherichia coli (except those known to be non-pathogenic)

Flavobacterium meningosepticum

Francisella tularensis (Type B)

Fusobacterium spp.

Gardnerella vaginalis

Haemophilus spp.

Hafnia alvei

Kingella kingae

Klebsiella spp.

FUNGI

Hazard Group 3

Blastomyces dermatitidis

(Ajellomyces dermatitidis)

Coccidioides immitis

Histoplasma capsulatum var. capsulatum

(Ajellomyces capsulata)

Hazard Group 2

Absidia corymbifera

Acremonium falciforme

Acremonium kiliense

Acremonium recifei

Aspergillus flavus

Aspergillus fumigatus

Aspergillus nidulans

Aspergillus niger

Aspergillus terreus

Basidiobolus haptosporus

Salmonella spp. (other than those in Hazard Group 3)

Serratia liquefaciens

Serratia marcescens

Shigella spp. (other than that in Hazard Group 3)

Staphylococcus aureus

Streptobacillus moniliformis

Streptococcus spp. (except those known to be non-pathogenic for Man)

Treponema pallidum

Treponema pertenue

Veillonella spp.

Vibrio cholerae (incl El Tor)

Vibrio parahaemolyticus

Vibrio spp. (other species known to be pathogenic for Man)

Yersinia enterocolitica

Yersinia pseudotuberculosis subsp pseudotuberculosis

Histoplasma capsulatum var duboisii

Histoplasma capsulatum var farciminosum

Paracoccidioides brasiliensis

Penicillium marneffei

Exophialia jeanselmei

Exophialia spinifera

Exophialia richardsiae

Fonsecaea compacta

Fonsecaea pedrosoi

Fusarium solani

Fusarium oxysporum

Geotrichum candidum

Hendersonula toruloidea

Leptosphaeria senegalensis

Hazard Group 2 (Cont)

Candida albicans

Candida glabrata

Candida guilliermondii

Candida krusei

Candida parapsilosis

Candida kefyr

Candida tropicalis

Cladosporium carrionii

Conidiobolus coronatus

Cryptococcus neoformans

(Filobasidiella neoformans)

Cunninghamella elegans

Curvularia lunata

Emmonsia parva

Emmonsia parva var. crescens

Epidermophyton floccosum

Exophialia dermitidis

Exophialia werneckii

Madurella mycetomatis

Madurella grisea

Malassezia furfur

Microsporum spp.

Neotestudina rosatii

Phialophora verrucosa

Piedraia hortae

Pneumocytis carinii

Pseudallescheria boydii

Pyrenochaeta romeroi

Rhizomucor pusillus

Rhizopus microsporus

Rhizopus oryzae

Sporothrix schenckii

Trichophyton spp.

Trichosporon beigelii

Xylohypha bantiana

PARASITES

Hazard Group 3

Echinococcus spp.

Leishmania spp. (mammalian)

Naegleria spp.

Toxoplasma gondii

Trypanosoma cruzi

Hazard Group 2

Acanthamoeba spp.

Ancylostoma duodenale

Angiostrongylus spp.

Ascaris lumbricoides

Babesia microti

Babesia divergens

Balantidium coli

Brugia spp.

Capillaria spp.

Clonorchis sinensis

Cryptosporidium spp.

Dipetalonema streptocerca

Dipetalonema perstans

Diphyllobothrium latum

Drancunculus medinensis

Loa loa

Mansonella ozzardi

Necator americanus

Onchocerca volvulus

Opisthorchis spp.

Paragonimus westermanni

Plasmodium spp. (human & simian)

Pneumocystis carinii

Schistosoma haematobium

Schistosoma intercalatum

Schistosoma japonicum

Schistosoma mansoni

Stronglyloides spp.

Taenia saginata

Taenia solium

Hazard Group 2 (Cont)

Entamoeba histolytica

Fasciola hepatica

Fasciola gigantea

Fasciolopsis buski

Giardia lamblia

Hymenolepis nana (human origin)

Hymenolepis diminuta

Toxocara canis

Trichinella spp.

Trichomonas vaginalis

Trichostrongylus spp.

Trichuris trichiura

Trypanosoma brucei subsp

Wuchereria bancrofttii

VIRUSES

Hazard Group 4

Arenaviridae

Junin virus

Lassa fever virus

Machupo virus

Mopeia virus

Bunyaviridae

Nairoviruses

Congo/Crimean haemorrhagic fever

Togaviridae

Flaviviruses

Tick-borne viruses

Filoviridae

Ebola virus

Marburg virus

Absettarov

Hanzalova

Hypr

Poxviridae

Variola (major & minor) virus

('whitepox' virus)

Kyasanur Forest

Omsk

Russian spring-summer encephalitis

Hazard Group 3

Arenaviridae

Lymphocytic choriomeningitis virus (LCM)

Rhabdoviridae

Rabies virus

Bunyaviridae

Bunyamwera supergroup

Oropouche virus

Phleboviruses

Rift Valley fever

Hantaviruses

Hantaan (Korean haemorrhagic fever)

Other hantaviruses

Togaviridae

Alphaviruses

Eastern equine

encephalomyelitis

Venezuelan equine

encephalomyelitis

Western equine

encephalomyelitis

Hepadnaviridae

Hepatitis B virus

Hepatitis B virus + Delta

Flaviviruses

Japanese B encephalitis

Kumlinge

Herpesviridae

Herpesvirus simiae (B virus)

Louping ill

Murray Valley encephalitis (Australia encephalitis)

Poxviridae

Monkeypox virus

Powassan

Rocio

Hazard Group 3 (Cont)

Retroviridae

Human immunodeficiency viruses (HIV)

Human T-cell lymphotropic viruses (HTLV) types 1 and 2

Hazard Group 2

Adenoviridae

Arenaviridae

other arenaviruses

Astroviridae

Bunyaviridae

Hazara virus

other bunyaviruses

Caliciviridae

Coronaviridae

Herpesviridae

Cytomegalovirus

Epstein-Barr virus

Herpes simplex viruses types 1 and 2

Herpesvirus varicella-zoster

Human B-lymphotropic virus (HBLV - human herpesvirus type 6)

Orthomyxoviridae

Influenza viruses types A, B & C

Influenza virus type A-recent isolates

Paramyxoviridae

Measles virus

Mumps virus

Newcastle disease virus

Parainfluenza viruses types 1 to 4

Respiratory syncytial virus

Papovaviridae

BK and JC viruses

Human papillomaviruses

Parvoviridae

Human parvovirus (B19)

St Louis encephalitis

Tick-borne encephalitis

Yellow fever

Picornaviridae

Acute haemorrhagic conjunctivitis virus (AHC)

Coxsackieviruses

Echoviruses

Hepatitis A virus (human enterovirus type 72)

Polioviruses

Rhinoviruses

Poxviridae

Cowpox virus

Molluscum contagiosum virus

Orf virus

Vaccinia virus

Reoviridae

Human rotaviruses

Orbiviruses

Reoviruses

Rhabdoviridae

Vesicular stomatitis virus

Togaviridae

Other alphaviruses

Other flaviviruses

Rubivirus (rubella)

Unclassified viruses

Hepatitis non-A non-B viruses

Norwalk-like group of small round structured viruses

Small round viruses (SRV - associated with gastroenteritis)

Unconventional agents associated with:

Creutzfeldt-Jakob disease

Gertsmann-Sträussler-Schienker syndrome

Kuru

Index

A

B

H

I

K

L

M